VALUE-DRIVEN PURCHASING

MANAGING THE KEY STEPS IN THE ACQUISITION PROCESS

THE NAPM PROFESSIONAL DEVELOPMENT SERIES

Michiel R. Leenders
Series Editor

Volume I
VALUE-DRIVEN PURCHASING
Managing the Key Steps in the Acquisition Process
Michiel R. Leenders
Anna E. Flynn

Volume II
MANAGING PURCHASING
Making the Supply Team Work
Kenneth H. Killen
John W. Kamauff

Volume III
VALUE-FOCUSED SUPPLY MANAGEMENT
Getting the Most Out of the Supply Function
Alan R. Raedels

Volume IV
PURCHASING FOR BOTTOM LINE IMPACT
Improving the Organization Through Strategic Procurement
Lisa M. Ellram
Laura M. Birou

VALUE-DRIVEN PURCHASING

MANAGING THE KEY STEPS IN THE ACQUISITION PROCESS

Volume I
The NAPM Professional Development Series

Michiel R. Leenders
Anna E. Flynn

**National
Association of
Purchasing
Management**

Tempe, Arizona

IRWIN
Professional Publishing
Burr Ridge, Illinois
New York, New York

Senior sponsoring editor: Cynthia A. Zigmund
Project editor: Jane Lightell
Production manager: Pat Frederickson
Designer: Larry J. Cope
Art manager: Kim Meriwether
Typeface: 11/13 Times Roman
Printer: Book Press

Library of Congress Cataloging-in-Publication Data

Leenders, Michiel R.

Value-driven purchasing: managing the key steps in the acquisition process / Michiel R. Leenders and Anna E. Flynn.

p. cm. — (The NAPM professional development series; v. 1)
"National Association of Purchasing Management."
Includes index.

ISBN 0–7863–0236–4
1. Industrial procurement — United States — Management. 2. Value analysis (Cost control) 3. Letting of contracts — United States.
I. Flynn, Anna E. II National Association of Purchasing Management.
III. Title. IV. Series.
HD39.5.L436 1995
658.7'2—dc20   94–21172

Printed in the United States of America
1 2 3 4 5 6 7 8 9 0 BP 1 0 9 8 7 6 5 4

SERIES OVERVIEW

The fundamental premise for this series of four textbooks is that effective purchasing or supply management can contribute significantly to organizational goals and strategies. This implies that suppliers and the way organizations relate to them are a major determinant of organizational success.

Differences do exist between public and private procurement; between purchasing for service organizations, manufacturers, retailers, distributors, and resource processors; between supplying projects, research and development, job shops, and small and large organizations across a host of industries, applications, and needs. Nevertheless, research has shown much commonality in the acquisition process and its management.

These four textbooks, therefore, cover the common ground of the purchasing field. They parallel the National Association of Purchasing Management (NAPM) Certification Program leading to the CPM designation. They also provide a sound, up-to-date perspective on the purchasing field for those who may not be interested in the CPM designation.

The textbooks are organized into the following four topics:

1. *Value-Driven Purchasing: Managing the Key Steps in the Acquisition Process*
2. *Managing Purchasing: Making the Supply Team Work*
3. *Value-Focused Supply Management: Getting the Most out of the Supply Function*
4. *Purchasing for Bottom-Line Impact: Improving the Organization through Strategic Procurement*

Volume I, *Value-Driven Purchasing: Managing the Key Steps in the Acquisition Process*, focuses on the standard acquisition process and its major steps, ranging from need recognition and purchase requests to supplier solicitation and analysis, negotiation, and contract execution, implementation, and administration.

Volume II, *Managing Purchasing: Making the Supply Team Work*, focuses on the administrative aspects of the purchasing department, including the development of goals and objectives, maintenance of files and records, budgeting, and evaluating performance. It also discusses the personnel issues of the function: organization, supervision and delegation of work, evaluating staff performance, training staff, and performance difficulties.

Volume III, *Value-Focused Supply Management: Getting the Most out of the Supply Function*, commences with identifying material flow activities and decisions, including transportation, packaging requirements, receiving, and interior materials handling. It goes on to cover inventory management and concludes with supply activities such as standardization, cost reduction, and material requirements planning.

Volume IV, *Purchasing for Bottom-Line Impact: Improving the Organization through Strategic Procurement*, begins with purchasing strategies and forecasting. This is followed by internal and external relationships, computerization, and environmental issues.

It is a unique pleasure to edit a series of textbooks like these with a fine group of authors who are thoroughly familiar with the theory and practice of supply management.

Michiel R. Leenders
Series Editor

ACKNOWLEDGMENTS

This textbook, like the other three in this series, was based on the sixth edition of the CPM Study Guide. This 1992 NAPM publication was intended to assist those preparing for the Certified Purchasing Manager (CPM) examinations. The Guide and its predecessors represented the collective work of a large number of purchasing academics and professionals who acted as editors, authors, and reviewers. For the sixth edition these tasks fell to Eugene Muller and Donald W. Dobler, editors; and Harry Robert Page and Eberhard E. Scheuing, consulting editors.

The authors were

Prabir K. Bagchi
Judith A. Baranoski
Lee Buddress
Joseph L. Cavinato
Michael J. Dunleavy
Donald J. Fesko
Barbara B. Friedman
Henry F. Garcia
Larry C. Giunipero
LeRoy H. Graw

Mary Lu Harding
Earl Hawkes
H. Ervin Lewis
Charles J. McDonald
Paul K. Moffat
Norbert J. Ore
Alan Raedels
Merle W. Roberts
Rene A. Yates

The editorial review board included

William F. Armstrong
John D. Cologna
Barbara Donnelly
Jack Livingston
Earl F. Pritchard

Karen Swinehart
Marvin C. Sanders
Archie J. Titzman
Earl Whitman

About one-third of their work has been used verbatim in this text. It has not been quoted in the traditional academic sense because it has been intermixed so thoroughly as to make clean separation almost impossible. The same approach has traditionally been used in the preparation of the Guide itself. We made the choice that, if no improvement was possible on what had previously been written, there was no point in rewriting it. We are, therefore, most grateful to the fine work of all CPM Guide contributors over the years.

New ideas in other management areas, as well as procurement, continue to expand the body of professional knowledge in our field. Thus, every attempt has been made to build on the solid foundation laid by many others in the CPM Study Guides and to bring it into relevance for today's environment. This has meant substantial reorganization and additions and a complete format change.

Every book is a collaborative effort. A number of people played a special role with respect to this text. Paul Novak at NAPM acted as the association anchor; Kathleen Little at NAPM served as the ultimate library resource. Jean Geracie at Irwin Professional Publishing got the project started and then passed the baton to Cynthia Zigmund for the next phases.

Sue LeMoine at the Western Business School not only helped in the production of the manuscript but was also full of suggestions for improvement. Her graphics skills have earned her a well-deserved reputation, and her contribution throughout this text is obvious.

We would also like to thank our families, friends, and colleagues for their forbearance during this literary effort. Unfortunately, no one has yet found a way of instant writing. Time devoted to this activity had to come from somewhere.

Michiel R. Leenders
Anna E. Flynn

CONTENTS

CHAPTER 1

INTRODUCTION

The fundamental premise for this textbook, as well as the other three in this series, is that effective purchasing or supply management can contribute significantly to organizational goals and strategies. This implies that suppliers and the way organizations relate to them are a major determinant of success.

This text focuses on the acquisition process, the sequence of steps necessary to ensure that an organizational need is satisfactorily met by one or more suppliers. In this introduction, the potential contribution of suppliers to organizational goals and strategies is discussed to put the acquisition process in a suitable context.

BETWEEN CUSTOMERS AND SUPPLIERS

If one views any organization as located between customers on the one hand and suppliers on the other (see Figure 1–1), it is not difficult to see that this is a continuous chain of organizations. Each organization places demands on its suppliers that are appropriate to meet its aspirations in terms of customer satisfaction (see Figure 1–2).

The supply chain can also be seen as equivalent to a typical input-transformation-output system. That the inputs should influence both the transformation process and the output (and vice versa) should be no surprise to anyone, for as the saying goes, "We are what we eat." That customer goodwill is seen as a major corporate asset is self-evident. What many organizations and companies are starting to recognize, however, is that supplier goodwill is also a key asset; and, in order to achieve

2

FIGURE 1-1
The Supplier-Customer Perspective of Supply

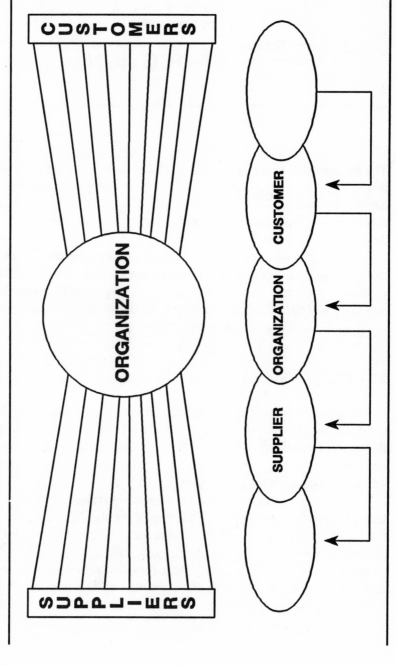

FIGURE 1–2
Total Customer Satisfaction Depends on Supplier Performance

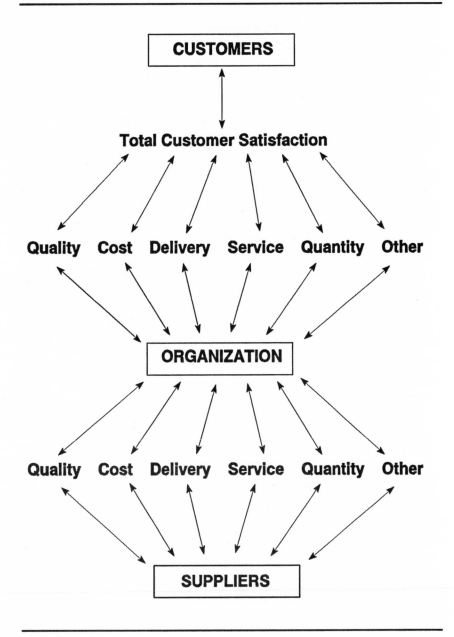

international accomplishment and competitiveness, companies and organizations must deal with their customers, their own employees, and their suppliers in a congruent and effective way. To ensure that the contribution potential of the supply function and suppliers is maximized requires a substantial management effort. This effort must come from people thoroughly familiar with the body of knowledge of the supply field, capable of identifying and exploiting opportunities that will ultimately increase their external customer satisfaction and their own organization's ability to prosper. This textbook and its three companions aspire to provide this vision as well as the means to accomplish it.

By definition, every organization plays three major roles in the supply chain. The first is as a customer to some supplying organization. The second is as a transformer of the input; and the third is as a supplier to another customer or organization (see Figure 1–3). This normally involves more than just three organizational entities because each organization in the chain continues to have customers and suppliers. For consumer products, the consumer is often identified as the "endcustomer," implying an end to the chain.

Within each organization a similar pattern may exist as information, materials, products, or services are passed from one department to another. The term *internal customers* is often used to distinguish them from *external customers*. The traditional notion has been that the supply function provides for the needs of such internal customers. However, there is growing concern that focusing exclusively on internal customers may not be in the best interest of external customers.

VALUE CHAIN

The term *value chain* refers to all of the transforming activities performed upon an input to provide value to a customer. The identification of this sequence of transactions is the first step in analyzing whether value is, in fact, added at every step and in finding better ways—better in quality, cost, timeliness, or value—to achieve customer satisfaction. Although in many organizations the value chain refers to internal operations, in a broader context, the value chain includes customers as well as suppliers

FIGURE 1–3
Input-Output Perspectives of Supply

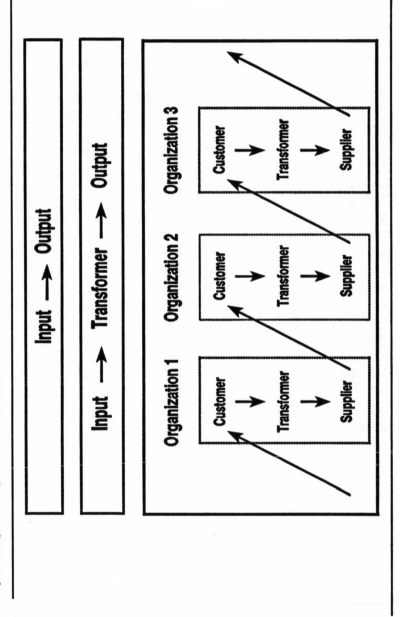

and their suppliers and so on. The larger challenge is to ensure that value is added all along the chain.

THE CONTRIBUTION OF PURCHASING AND SUPPLY

The contribution of the purchasing and supply function can be highlighted in a number of ways. The first focuses on the operational and the strategic dimensions (see Figure 1-4).

The Operational Contribution

Traditionally, there has been a strong emphasis on the operational contribution of the purchasing function. This emphasis may be summarized by the classic phrase, "ensuring that the right goods and services, of the right quality, arrive at the right time, in the right place, in the right quantities, at the right price." For decades this has been described as the key objective of the purchasing function. Taken literally, this objective has both an operational and a strategic dimension. Unfortunately, it has taken on the connotation of being primarily operational. An internal customer identifies a need and transmits a purchase request to the purchaser, who proceeds to fill it. If this need is not properly filled, serious consequences may arise for both the internal and external customer. Inferior quality, late delivery, and insufficient or excessive quantities can be anything from a minor nuisance to a very major concern. Should difficulties arise, immediate attention is required to find a solution. The operational side of purchasing is heavily focused on ensuring that needs are properly met, thereby preventing and solving problems. Given the wide range of needs of most organizations, plus the requirement for a documentation trail (after all, organizational funds are involved), the operating supply task is composed of a continuous series of transactions to ensure that the organization can continue to function daily. Over time, the pressure of daily needs and constant attention has forced the function into a highly reactive mode. In fact, most purchasers see themselves as a service function, primarily concerned with internal customers and short-term needs. They see themselves as overworked and underappreciated.

FIGURE 1-4
The Three Perspectives on Supply Contribution to Organizational Objectives and Strategies

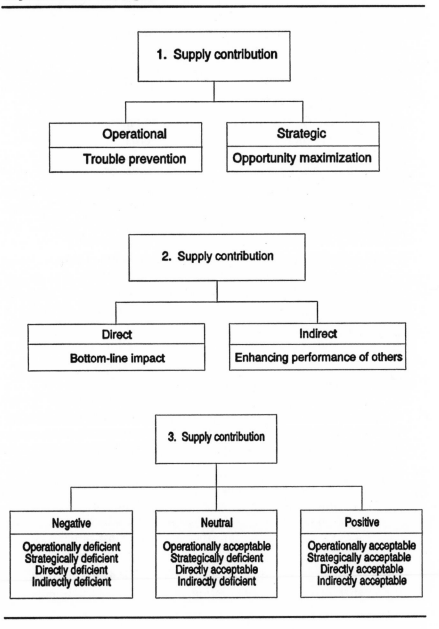

In a sports analogy, purchasers, as well as their internal customers, see themselves as providing uniforms, towels, supplies, and drinks to the team. When the game is on, they sit on the sideline and watch the team players do the scoring. This is a highly destructive view of the purchasing function and its potential contribution. It is further reinforced by the accounting view of organizational responsibilities and delegation. Accountants have long maintained that fraud could be minimized if different groups of employees specified requirements, used them, sourced them, and paid for them. Therefore, the reason for a supply function is a negative one—employees cannot be trusted around money. Since no value is added by the prevention of fraud from the external or even the internal customer's perspective, the existence of a supply department is seen as a bureaucratic necessity. Obviously, the less that is spent to operate such a department, the better. In this context, the supply function is seen as an idler gear in the organizational transmission. Organizational design engineers have not quite found a way to get rid of it yet, but they are working on it with reengineering and computers.

The operational contribution of supply is more than just crime prevention. Assuring that daily needs are met properly requires proper attention and specialized supply knowledge and skills. The foremost responsibility of supply is to ensure that the ongoing operational needs of the organization are met.

The Strategic Contribution of Supply

The strategic contribution of supply has a longer-term, broader perspective focused on organizational goals and strategies and the external customer's satisfaction. Because the supply function is simultaneously exposed to the needs of the organization on one hand and the market on the other, it is in a unique position to identify and exploit opportunities not apparent to others in the organization (see Figures 1–4 and 1–5). The strategic contribution of supply requires the function to ensure that suppliers, and the way the organization relates to them, become and remain a source of competitive advantage.

The strategic perspective on the supply function sees it as part of the engine that drives the organization to future success. It puts the

Figure 1-5
The Strategic Contribution of Supply

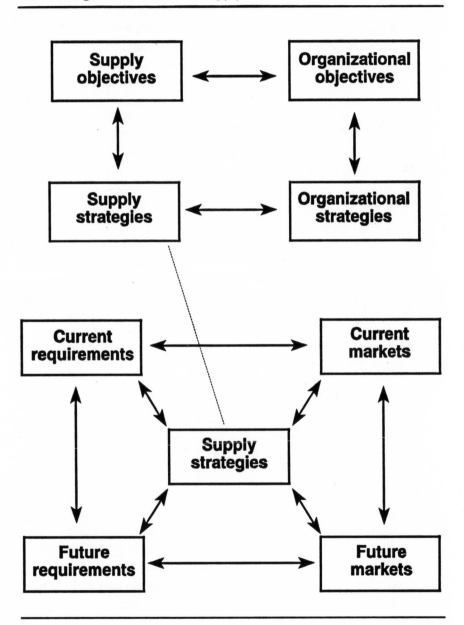

purchasing function and suppliers directly on the team as players instead of on the sideline. It requires a longer-term, much more aggressive perspective of the purchasing function and of suppliers. It requires that organizational objectives and strategies incorporate supply considerations and opportunities. That is why in Figure 1–5 the arrows are shown as going both ways. Supply objectives and strategies should be congruent with organizational objectives and strategies. Equally important, organizational objectives and strategies should appropriately reflect supply opportunities as well as supply constraints.

That organizational objectives and strategies cannot be set independently of the market on the supply side should not be such a novel idea. After all, most organizations recognize their dependence on the external customer market. Unfortunately, the prevalent management notion that whatever is needed can always be procured has obscured the need for careful supply market assessment along with customer market evaluation. The reality of this need dawns when significant shortages or change such as military conflict, OPEC restrictions, or agricultural disasters occur in supply markets.

For organizations that purchase significant amounts of commodities, price and availability fluctuations can represent significant risks. For a candy bar manufacturer, for example, fluctuations in the price of cocoa and sugar may well represent one of the largest business risks to which the organization is exposed.

New technology represents another area of opportunity and risk on the supply side because of its rapid introduction and potential for making previous supply solutions obsolete. Purchasers should be sensitive to news about new technology in the market, be able to assess its potential impact, and ensure its conveyance to suppliers and internal and external customers where appropriate.

The strategic perspective on supply goes beyond operational needs into risk management, technology, competitive intelligence, and continuous improvement. It requires that top management be significantly involved in the creation and execution of supply strategies and also be in direct contact with the top management of key suppliers. The strategic dimension of supply is further explored in the fourth textbook in this series, but it should be kept in mind throughout all texts.

The Direct versus the Indirect Contribution of Supply

A different perspective of supply and suppliers comes from the direct versus the indirect contribution (see Figure 1–4). The direct contribution of supply is that purchasing savings flow directly to the bottom line. If it is possible to secure a lower price for exactly the same quantity of exactly the same requirement or to find a lower-cost substitute that performs essentially the same function, dollars will be saved. Since, for most organizations, purchases of goods and services range from 30 to 80 percent of organizational revenue, such savings may significantly affect the bottom line. In public institutions such savings may reduce deficits, contribute to surpluses, or allow the organizations to hold the line on taxes or prices charged for services or even to extend services. In private organizations, such savings can have a major effect on profitability. For example, take a manufacturer that spends 65 percent of revenue on outside goods and services, 20 percent on labor and administration, and 10 percent on financial charges and has a 5 percent profit before tax. A 4 percent direct saving in purchasing could increase overall profitability to 7.6 percent for a 52 percent increase!

Notably, a 4 percent purchasing savings target is not at all unreasonable. For an organization like the manufacturer mentioned to achieve an equivalent profit increase from labor and administration costs would require a 13 percent decrease, far more difficult to achieve. The alternative of increasing sales by at least 40 percent might be equally difficult.

This argument has traditionally been most prominent in advocating the need for the purchasing function. Savings generated normally exceed the cost of running the department, so the function pays for itself, and any extra benefits come gratis. As more organizations downsize and outsource, and as the percentage of revenue spent on labor and administration decreases, the potential bottom-line impact of purchasing savings continues to increase.

Supply also contributes indirectly to the bottom line by enhancing the performance of other departments or individuals in the organization. For example, better quality may reduce rework, lower warranty costs, increase customer satisfaction, or increase the ability to sell more or at a higher price. All of these benefits will ultimately affect the bottom line, but less directly. Faster delivery may reduce inventories, allow for quicker introduction of new products or services, increase flexibility, or gain market share, to name just a few benefits. What is interesting about the indirect contribution of supply and suppliers is that preliminary research shows that its benefits may well outweigh the direct, traditional bottom-line impact. If this is true, greater focus on this aspect of supply contribution should have tremendous payoff.

> One of the advantages of teams is the expertise brought to the team by each member. For purchasing, an added advantage may be a greater realization on the part of other groups of the importance of purchasing. Also, problems or delays may be detected before they occur, thus speeding up the total process in the long run.
> In smaller to midsize public purchasing organizations where departments are small, it may be impossible for a purchaser to know everything there is to know about all the different goods and services he or she buys. The director of procurement for the City of Tucson, Arizona, worked on a cross-functional team to draft a Procurement Code that was passed by the City Council. Washoe Medical Center in Reno, Nevada, has an ongoing product standards team designed to support Washoe's quality efforts. McCarren International Airport in Las Vegas, Nevada, uses a cross-functional team which includes the purchasing and contract administrator to oversee emergency management at the airport.[1]

In today's environment of relatively low inflation, more and more purchasers are trying to get suppliers to keep prices stable or actually lower prices. It is equally important that purchasers put pressure on suppliers for improved value in the nonprice dimensions of their offerings. At Corning, for example, suppliers are asked to submit ideas for changes and to prove that their ideas will result in measurable value improvement.

---

[1]"Teaming in Public Purchasing," *NAPM Insights*, March 1993, pp. 34-35.

Negative, Neutral, and Positive Contribution

Another perspective on the contribution of supply to organizational objectives and strategies can be obtained from suppliers themselves as well as internal and external customers. Surveying or benchmarking their perceptions on the contribution of supply and suppliers can provide valuable insight into the role of supply in an organization.

Negative Contribution

General experience shows that asking suppliers and internal and external customers about supply is likely to result in a range of reactions. A reaction is deemed to be negative if the responder feels he or she has difficulty performing his or her job well because of supply difficulties. These individual findings can be summarized by department or for an individual customer or group of customers. A negative contribution would mean that the operational side is deficient, as are all other dimensions of the potential contribution (see Figure 1–4). An aggregation of internal and external customers' responses resulting in an overall negative perception of supply is very serious. It requires significant management action, often including the removal of key supply personnel and their replacement with more competent and compatible individuals. Incidentally, most supply groups tend to overestimate their contribution compared to the ratings by customers or their fellow employees in other departments.

Neutral Contribution

If the overall ranking is neutral, at least the operational side is under control. Suppliers, internal and external customers, and supply managers do not see supply as a significant potential contributor to organizational goals and strategies or as much of an indirect potential. Top management involvement is just sufficient to prevent supply problems.

Positive Contribution

An overall positive contribution means that suppliers and internal and external customers are aware that supply and suppliers allow the organization to provide customer satisfaction beyond what other similar organizations can achieve. Because of the kinds of suppliers the organization has and the way it relates to them, it can outperform others. In this context, supply contribution is positive both operationally and strategically, as well as directly and indirectly.

This positive contribution is difficult to attain and maintain. It requires significant top management involvement in supply as well as the active support of other functional groups and customers. In the sports analogy, it requires that supply and suppliers are on the field or on the court and play right alongside the players. To illustrate the difficulties of accomplishing this feat effectively, imagine the coach of a football team trying to win with a team composed of some exclusive players, others who are nonexclusive, and a significant number who excel at other sports. Yet, David N. Burt and Michael F. Doyle argue that it is exactly this kind of team that will allow organizations to prosper.[2]

Difficult as the positive contribution may be to attain, it does represent the professional aspiration of those engaged in the supply field, whether or not competitive or economic pressure requires it. Positive contribution is the goal pursued in this and the other three textbooks in this series.

Short- and Long-Term Considerations

Traditionally, supply decisions have been relatively short term, paralleling the operational focus. Short-term flexibility is often prized in the pursuit of customer satisfaction or competitiveness. Strategic and indirect contributions frequently require a longer time. However, the idea of investing now to reap benefits later may be difficult to sell in organizations under significant financial pressure. Nevertheless, the key questions need to be asked: What are the long-term goals and strategies for this organization and what contribution is required from supply and suppliers to help achieve them? How should organizational goals and strategies be changed to reflect opportunities in the supply function and the market?

---

[2]David N. Burt and Michael F. Doyle. *The American Keiretsu* (Homewood, Illinois: Irwin, 1993).

Public and Private Supply

Every organization, public or private, needs suppliers to help it accomplish its mission. No organization can survive without suppliers. Public and private organizations share concern for value of funds expended. However, public procurement normally requires that supply be managed within the context of a legal structure binding the purchaser to certain policies and procedures designed to ensure accessibility and fairness to suppliers as well as auditability and openness to the taxpayer. Public purchasers are often required to advertise their requirements and may be restricted on the length of contract they are allowed to engage in. Public purchasers have requirements with respect to GATT, minority or woman-owned business, local content, bid and performance bonds, and so on. Every effort will be made in this and the other three textbooks to highlight significant differences between public and private practice where appropriate.

COMPUTERIZATION

The large amount of administrative detail and high number of transactions that are part of the supply function make it almost imperative that computerized systems be used extensively. Even though computerization and electronic data interchange (EDI) will be mentioned in each of the first three textbooks where appropriate, primary coverage will be provided as part of the fourth textbook on purchasing strategy.

ETHICS

Ethical conduct in supply is essential. Large sums of money and confidential information are frequently involved. Suppliers use some rather standard tools of persuasion and also develop innovative approaches to induce purchasers to buy. Internal interactions between purchasers and users, specifiers, requisitioners, inspectors, receivers, and payers need to be conducted properly to ensure effective and efficient supply management. Some organizations insist that every employee be able to pass the mirror test—can you honestly face yourself daily in the mirror and defend every decision and action taken? The National Association of Purchasing Management (NAPM) has developed standards and guidelines for ethical purchasing practice (domestic and international), which provide ethical guidance for the supply function. These are discussed in Chapter 7 of this book. In the following descriptions of the acquisitions process and its various decisions, the ethical considerations put forth by NAPM will be assumed in force throughout. Honesty, fairness, and trust have to be the driving values for effective supply management.

KEY STEPS IN THE ACQUISITION PROCESS

The heart of this book is devoted to the acquisition process, the series of steps in which a need is translated into a requirement and ultimately satisfied with the assistance of a supplier. It is appropriate that the first book in this series covers the acquisition process, given that a fundamental understanding of that process should be a prerequisite for each of the three remaining books.

The key steps in the acquisition process for a new requirement are:

1. Determination of need.
2. Communication of need.
3. Identification of potential sources.
4. Solicitation and evaluation of bids and proposals.
5. Preparation of the purchase order.
6. Follow-up and expediting.
7. Receipt and inspection.
8. Clearing the invoice and payment.
9. Maintenance of records.

Not all of these steps require one full chapter in this text. Moreover, negotiation is such a significant activity in the buying and selling process that it has been given a chapter of its own. The following paragraphs describe the coverage of each chapter.

Chapter 2 covers the determination of need and provides a classification of purchases. It discusses the internal communication of needs, requisitions and their review, and conformance to organizational, legal, and environmental and ethical requirements. Specifications conclude this chapter.

Chapter 3 is concerned with potential supply sources and supplier selection and evaluation considerations. Chapter 4 covers solicitation of bids, proposals, and information. Chapter 5 deals with bid analysis, price and cost analysis, and the decision to make or buy.

Chapter 6 is devoted to negotiation. It describes reasons for negotiation, the negotiation process, the negotiation plan, and the various tools and techniques appropriate to the steps in the negotiation process.

Chapter 7 covers the execution, implementation, and administration of the contract, follow-up and expediting, receipt of goods, and payment of suppliers. NAPM's standards of conduct and ethics are the last topic covered in this chapter. Chapter 8 discusses legal issues in supply.

KEY POINTS

1. The supply function can contribute significantly to organizational goals and strategies.

2. Supply contribution is operational and strategic; direct and indirect; negative, neutral, and positive.

3. Value-focused customer satisfaction, quality, continuous improvement, international competitiveness, time compression, and lowest cost of ownership require that supply shed its service image and that suppliers and supply become full team players.

4. Every organization is part of a value chain in which it simultaneously performs the roles of customer, transformer, and supplier.

5. The first in the series of four texts, this book focuses on the acquisition process. Book 2 deals with the administration of the supply function, Book 3 on supply and inventory management, and Book 4 on purchasing strategy and current issues.

6. Proper ethical conduct is vital to effective supply management.

CHAPTER 2

DETERMINATION OF NEEDS

Recognizing the needs of the organization is the first step in the acquisition process. Throughout this process, and particularly during the needs identification step, it is important to remember the dual contribution of supply to organizational objectives described in Chapter 1. The purchaser must perform the operational aspects effectively and efficiently while mindful of long-term goals and opportunities for the organization. This chapter discusses in detail the operational (or trouble avoidance) side of supply management. The purchaser must balance these operational tasks with both the strategic, opportunity-maximizing side of supply management and the indirect contribution of enhancing the performance of others.

While this responsibility for recognition and determination of need lies mainly with the person accountable for a particular activity, the purchasing manager or buyer should help anticipate the needs of the requisitioner—what is needed, when, how much, and what quality.

CLASSIFICATION OF NEEDS

Faced with the pressures of global competition, the demand for continuous improvement, and a renewed emphasis on customer, employee, and supplier satisfaction, every organization must continuously improve materials management decisions.

Supply management must ensure that organizational needs are effectively met in both the short and long term. Organizational needs cover a wide range. A classification of these needs is useful to ensure that each is properly met and that value is received. Four major classifications include: (1) type of need, (2) strategic or operational, (3) repetitive or nonrepetitive, (4) ABC (or monetary value).

Types of Needs

Needs may be categorized or niched as:

1. Raw materials (chemicals, metals).
2. Purchased parts (components like semiconductor chips and semifinished goods like iron castings).
3. Packaging (paper, plastic, glass, corrugated).
4. MRO (maintenance, repair, and operating supplies like paint, spare parts, lubricants).
5. Services (janitorial, security, consulting).
6. Capital equipment (machinery, office equipment, vehicles).
7. Resale (finished goods, like private label audio systems).

Each category has its own technologies, supply base, logistics systems, cost and price characteristics, market conditions, and trade practices. Thorough familiarity with these trade parameters is a fundamental prerequisite of a good buyer.

Strategic or Operational

A requirement may be relatively low in value but may still have strategic significance for the organization. A good example might be a unique raw material that in small quantities enhances the performance of a much larger batch of raw material or of a piece of machinery. It is the spicing that gives Kentucky Fried Chicken its strategic advantage. Major dollar requirements are often strategic because of their bottom-line impact. Any requirement in tight supply may represent an opportunity for gaining competitive advantage. The strategic-operational division will indicate how much attention needs to be given to a particular supply requirement. Moreover, strategic requirements may involve secrecy, unusual speed, or unusual supplier involvement. These strategic needs probably mean that standard procurement methods, like inviting many suppliers to bid, may not be appropriate.

Repetitive or Nonrepetitive

Purchases may be classified depending on whether or not they are repetitive. Repetitive purchases should be made in a repeatable system of acquisition, such as blanket orders or systems contracting, which are described later. Nonrepetitive purchases should be dealt with differently.

It is wise to invest time and energy in the acquisition process for repetitive purchases. A suitable trade-off position needs to be found between the cost of acquisition and ensuring that the best value is obtained. Moreover, the repetitive nature of the requirement allows for review, process refinement, and continuous improvement. For nonrepetitive purchases this may be more difficult.

ABC Analysis

One system of classification that is useful in various areas of materials management is based on the monetary value of each type of purchase. Referred to as ABC analysis, the Pareto curve, or the 80-20 rule, this concept classifies inventory items, purchased items, or suppliers in descending order of dollars spent per year. The ABC breakdown results in three classes, A, B, and C, as follows:

Class	Percentage of Total Items Purchased	Percentage of Total Purchase Dollars
A items	10	70-80
B items	10-20	10-15
C items	70-80	10-20

A similar analysis can be done for ABC suppliers or ABC inventory items with similar results. Once classification is complete, an analysis compares a specific purchase to the percentage of total purchase dollars, or the number of suppliers per item to the dollar value of the items purchased. This type of analysis shows where to direct managerial time, effort, and funds. Class A purchases would be given more attention than B or C requirements. For a candy bar manufacturer the A requirements, cocoa, sugar, and packaging, would receive the most managerial attention and the greatest allocation of resources.

In operational terms, the goal for A, B, or C requirements is to ensure "the right good or service, right place, right time, right quality, right quantity, right price." This goal is achieved differently for A, B, and C requirements. Supply assurance, quality, and delivery may be as important for B and C requirements as for A requirements but the purchaser would manage these requirements differently. By using procurement methods such as blanket orders, systems contracting, consignment purchasing, supplier stocking programs, supplier-operated stores, and programmed releases against a schedule, the purchaser reduces paperwork, personnel, and time devoted to B or C requirements. Effectively implemented, ABC analysis may free up the time a purchaser needs to focus on the strategic aspects of the function.

The strategic focus on the supply side would normally be on the A requirements, which represent the most critical purchases for the achievement of organizational goals and objectives. Class A requirements represent the greatest opportunity for improving direct and indirect bottom-line impact. For A requirements price deserves careful scrutiny; the buyer must be constantly alert to market opportunities, substitutes, and new technologies to ensure the best value possible. Likewise, special delivery arrangements may be made for A requirements because the cost of carrying inventory may be too large a drain on working capital. Though often of less strategic importance than A requirements, B requirements still have significant financial impact. They lend themselves well to standard inventory and ordering systems. C requirements are also important to the continuing function of the organization, so availability must be assured. However, a number of C requirements may be safely stocked in sufficient quantities to avoid worrying about running out.

If the ABC concept is combined with type of purchase and whether acquisition is strategic or repetitive, a three-dimensional block is created. Figure 2–1 demonstrates the 84 different subsets created if the type of purchase is put on the X-axis, the ABC classification is put on the Y-axis, and the strategic-operational and repetitive-nonrepetitive division put on the Z-axis. Each subset requires specialized purchasing treatment.

FIGURE 2-1
Organization of Purchases

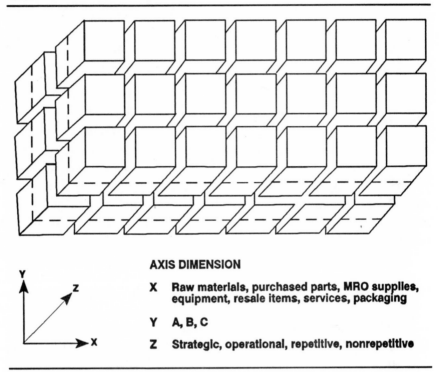

AXIS DIMENSION

X **Raw materials, purchased parts, MRO supplies, equipment, resale items, services, packaging**

Y **A, B, C**

Z **Strategic, operational, repetitive, nonrepetitive**

Further Classification Implications

Classification has many implications not yet mentioned. For example, an A requirement purchased repetitively in the purchased part category would require the assignment of a skilled and experienced buyer with good analytical and negotiation skills. The commitment of resources for a detailed and expensive review of suppliers might have a tremendous direct and indirect bottom-line impact.

The organization of the supply department may also be broken down along these same lines with buyers specializing in one of these categories. Expertise is required for each type of purchasing. For example, MRO supplies may be assigned to a relatively inexperienced buyer because many small-value items are ordered repeatedly and are often sole sourced. Even within this category, repair parts for production equipment may require more specialized knowledge. Raw materials purchases are likely to be assigned to an experienced buyer because raw material supply often involves large-volume contracts, a thorough knowledge of primary processes, and a high sensitivity to economics and the world supply situation.

Opportunity to Affect Value

The opportunity to add value within the acquisition process is generally greatest during the need recognition and description stages (see Figure 2–2). Therefore, early purchasing involvement and early supplier involvement during these stages can contribute greatly to value improvement. This point should be kept in mind throughout the remainder of this book.

PURCHASE REQUISITIONS

The traditional method of communicating a need from the requesting department to purchasing is through a purchase requisition (PR). There are two basic types of requisitions: the standard (single-use) requisition and the traveling requisition.

Standard Requisition

The standard purchase requisition is a form used to communicate the specific needs from the user department to purchasing and to the supplier. A requisition typically includes:

- The date.
- An identification number.
- The originating department.
- The account to be charged.

FIGURE 2-2
Opportunity to Affect Value During the Six Steps of the Acquisition Process

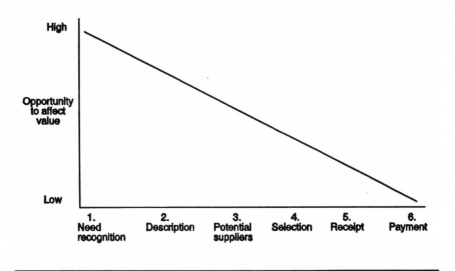

Source: Michiel R. Leenders and Harold E. Fearon, *Purchasing and Materials Management,* 10th ed. (Homewood, Illinois: Irwin, 1993), p. 137.

- A complete description of the desired material, part, service, or equipment.
- The quantity needed.
- The date needed.
- Any special shipping instructions.
- The signature of the authorized requisitioner.

The required information and the number of copies of the requisition vary with organizational needs. At a minimum, in a manual system purchasing and the user department retain a copy.

Traveling Requisition

Recurring requirements for materials and standard parts are often purchased via a traveling requisition. In a manual purchasing system, the requisition is a card containing standard part or item information along with listings of past requisitions and past suppliers. The requisition is sent to purchasing with an indication of the amount and date needed. Purchasing completes the ordering process and returns the traveling requisition with supplier information, the price, and a purchase order number. The user department stores the traveling requisition until it is time to reorder. As many as 24 to 36 purchases may be initiated from one card.

The traveling requisition reduces operating expenses because of the savings in paperwork and clerical time. The traveller also provides a complete, cumulative purchase history and use record on one form.

In a computerized system buyer and supplier are linked by computer; requests for goods and services and acknowledgments of action are transmitted by wire and screen. This eliminates the need for paper-based systems, reducing paper and clerical time, and often enhances decision processes and record keeping.

Bill of Materials Requisition

The bill of material (B/M) is used when standard manufactured items are produced over a long period. The B/M simplifies the requisitioning process for frequently needed items. The B/M lists the total number of parts or materials to make one end unit, including scrap allowance. Production scheduling notifies purchasing of the total number of end units scheduled for that model next month. Purchasing "explodes" the bill of material by multiplying to determine the total quantity of material needed to meet the next month's production schedule.

A comparison of needs to inventory reveals the open-to-buy amounts. In a fully integrated materials computer system, where supplier information and pricing are already in the database, the order releases are generated by the computer to cover the open-to-buy amounts.

Requisition Review

Purchase requisitions should be reviewed to determine their fit with organizational requirements. Ideally, the needs identification process continually compares purchase needs to the short- and long-term goals of the organization. The requisition process provides one more chance to review the decision to buy.

Four areas should be considered at this stage of the purchasing cycle: organizational, operational, financial, and marketing strategies.

Organizational Strategies
Purchasing has a unique position in the organization because of its exposure to internal and external environments. Purchasing can contribute greatly to the firm's competitive position in terms of short-term profits and long-term corporate growth. Developing buying plans congruent with the annual needs and the longer-term goals of the organization enables purchasing to provide direct input into the organization's strategic plan.

Operational Strategies
In a manufacturing or service environment, one overriding concern is to keep operations running smoothly. Purchasing plays a key role in the operational side of the house, and timing and coordination between purchasing and operations is crucial. If purchasing is given insufficient time to obtain a needed item or service, the organization may sacrifice efficiency, competitive position, or a negotiation edge. The end result of purchasing inefficiency is often higher prices, special production runs, or premium transportation costs. Late deliveries will reduce customer satisfaction.

Financial Strategies
Economic factors may have a positive or negative impact on buying opportunities. An organization may decide to buy in advance of actual needs to avoid higher prices later or, conversely, may delay a purchase if a price reduction is predicted. Financial forecasts and trends provide purchasing with the information necessary to make the greatest bottom-

line contribution to the organization and assist in determining the appropriate exposure.

Marketing Strategies

Marketing strategies should be driven by the firm's or organization's strategic plan. The sales forecasts become the basis for production schedules, which become the basis for purchasing decisions. Capital equipment budgets and other sales activities are influenced by sales forecasts.

CONFORMANCE TO CORPORATE POLICY, LEGAL, AND ETHICAL REQUIREMENTS

Financial Review

Expenditures may be budgeted or nonbudgeted. Ordinarily, expenditures are first authorized in the requisitioner's or organization's budget. The buyer is then responsible for bringing goods and services into the organization within the budgeted amount or less. Anticipated variances should be brought to the attention of the requisitioner prior to placement. Nonbudgeted items require close scrutiny and authorization prior to purchase.

Requisitions should also be reviewed against the budget to ensure availability of funds. Funds are usually made available through one of the following types of budgets:

Material budget—covers production materials and components.

Capital budget—covers the acquisition of equipment that is capitalized as a depreciable asset on a company's balance sheet.

MRO budget—covers items that are needed for the operation of a facility but are not part of the finished product.

Sometimes a purchase cost may be directly reimbursable under a contract. In government contracting, not all costs are recoverable or allowable by the cost accounting standards. Unallowable costs are identified in the Federal Acquisition Regulations (FAR).

In many public organizations purchasing is expected to ensure that the requisitioner has sufficient funds available to cover the purchase. In these organizations purchasing is part of the internal financial control system and plays a policing role that requisitioners may resent.

Cost Accounting Structure

An accounting system is the entire structure of records and procedures that discover, record, classify, and report information on the financial position and operations of an organization. For most organizations, the budget represents the authority for making expenditures. Up-to-date, adequate information therefore becomes critical.

To ensure adequate internal control for budgetary and accounting purposes, organizations typically have the following rules and procedures in place:

1. Purchases should be executed through purchase orders initiated by those with the proper authority.

2. Purchasing practices should be reviewed periodically to disclose inequities.

3. Incoming merchandise should be controlled through a receiving report.

4. Invoices should be forwarded directly to the accounting department for verification against the purchase order and receiving report.

5. Invoices should not be recorded as a liability until properly approved.

6. Accounting, purchasing, and receiving responsibility and activities should be adequately separated.

7. Suppliers' claims should be paid promptly.

These rules and procedures are often challenged when an organization tries to reengineer its processes. Additional control steps in the acquisition procedure may be costly, delay delivery, and add little value.

Health and Safety Laws

A number of laws and regulations related to health and safety and the environment affect procurement. Since the 1960s the government has dictated worker safety, particularly through the Occupational Health and Safety Act (OSHA). Buyers have attempted to protect themselves and shift responsibility to suppliers by including an OSHA compliance clause in their contracts.

The Consumer Product Safety Act gives a commission broad regulatory power to safeguard consumers against unsafe products. Purchasers have the responsibility to make sure the products they buy meet the requirements of the legislation.

The Environment

The Environmental Protection Agency (EPA) was established to implement and enforce federal laws relating to clean water and air, waste disposal, and related matters. Growing concern for hazardous waste disposal has also led to many changes in processes and products. As a result, purchasers and suppliers have had to change products, services, and methods of doing business.

Purchasers must be concerned about what they buy and, after usage, how waste will be disposed of. Throughout the acquisition process, purchasers must think of the total cost of using a specific product or service. Hazardous material disposal expenditures can quickly drive up the total cost, and motivate the purchaser to seek alternatives.

For a more in-depth discussion of environmental issues, refer to Book 4 of this series.

Confidentiality

According to the National Association of Purchasing Management (NAPM) Principles and Standards of Purchasing Practice, confidential or proprietary information belonging to employers or suppliers should be handled with due care and proper consideration of legal and ethical ramifications and governmental regulations.

Confidentiality is a critical issue for the integrity and professionalism of the buyer. Purchasers should realize that certain information should be kept strictly within the organization. Information about one supplier should never be divulged to a competing supplier.

Proprietary information requires protection of the name, composition, process of manufacture, or rights to unique or exclusive information that has marketable value and is upheld by patent, copyright, or nondisclosure agreement.

The following information may be considered confidential or proprietary according to the NAPM Guidelines:

- Pricing.
- Bid or quotation information.
- Cost sheets.
- Formulas and/or process information.
- Design information (drawings and blueprints).
- Company plans, goals, and strategies.
- Profit information.
- Asset information.
- Wage and salary scales.
- Personal information about employees or trustees.
- Supply sources or supplier information.
- Customer lists or customer information.
- Computer software programs.

NAPM suggests some general guidelines for dealing with confidential information. If a formal policy does not exist, purchasing should work with legal counsel to develop one. Divulge information within the organization on a need-to-know basis only. Clearly label information as confidential when transmitting it. Consider using disclosure and nondisclosure agreements to clarify parameters for use and responsibility of the information. All information should be protected and the effects of its use or disbursement considered.

Corporate Authorization

Most organizations have policies and procedures to ensure that necessary approvals have been obtained. Typically, the director of purchasing and delegated individuals are responsible for the commitment of funds to acquire materials, equipment, and services. Corporatewide contracts usually require the approval of the chief purchasing officer.

For standard materials in common and repetitive use, the purchasing manager has some latitude over purchasing for stock or in advance of specific requirements. The driver behind these purchases is typically a bill of material, a specific order from an internal customer, or a production plan.

For purchases in excess of a specified dollar amount, a policy usually requires proper, formal source evaluations. In the case of single or sole sourcing, buyers are usually required to document purchases and may attempt to develop other options. In an emergency situation, competitive bidding may be waived with appropriate management documentation. In the public sector, policies are less flexible, and more extensive use of formal advertising and competitive bidding is often required.

To maintain control over spending and guard against internal fraud, individuals are given specific dollar limits, called limits of authority or spending limits or delegations. Thus, a buyer may be able to place an order up to $10,000 and a senior buyer $25,000. In most organizations, the finance or accounting department maintains lists of authorized requisitioners and their respective signatory dollar limits. The purchaser must verify the approvals prior to placing an order. Incorrect requisitions are returned to the user. Often blanket approval is given for direct material items used in production and requested by inventory or production control. Contracts for continuing needs, such as blanket orders or open-end orders where the dollar expenditure level exceeds the individual's authority, should contain the manager's signature. This applies even though the individual order releases will be less than the buyer's authority limit.

SPECIFICATION OF NEED

An accurate description of the need, service, or good is essential to an effective and efficient procurement system. The buyer and the user share the responsibility for developing the appropriate specifications, performance terms, and acceptance criteria. The integrative nature of the business environment often means the buyer is part of a cross-functional team composed of the user, operations, quality, design, and finance. Whether the team is formal or informal, the buyer must have four rights or prerogatives if "good" purchasing is to follow. These rights include meaningful involvement in:

- Determination of the appropriate specifications.
- Supplier selection.
- Determination of the appropriate pricing method.
- Contact with potential and existing suppliers.

How these rights are exercised within the team will have a huge impact on the ability of the team members to work together and on the quality of specification.

The term *meaningful involvement* has been chosen to underline the idea that purchasing must add value through its involvement. The nature of the particular purchase and the expertise and skill of the purchaser—as well as of the requisitioner, specifier, and other members of the team—will determine the extent of purchasing contribution possible. Ian Stuart, in his doctoral dissertation, found that wide variations existed in the extent of purchasing involvement in capital acquisitions for R&D projects.[1]

The purpose of defining specifications, performance terms, and criteria is twofold:

1. To focus the thinking of the group that will eventually be using the products or services to ensure that the specifications, performance

---

[1]Stuart, F. Ian. *The Acquisition of Equipment in a Research and Development Environment*, doctoral thesis, Western Business School, 1989.

terms, and acceptance criteria have been fully developed. Looked at from a value-added standpoint, this step is critical. Without it, the buying company risks purchasing goods and services that do not meet its requirements or incurring unanticipated costs. Defining specifications up front can minimize excess costs.

2. To communicate the complete requirements of the transaction clearly to all participating suppliers. This ensures that all competitors quote to the same baseline (which will assist in a comparison of quotations) and that the quotations received address specifically the products and services requested.

Description Methods[2]

There are several methods of describing a required item to a supplier, including:

1. By brand.
2. "Or equal."
3. By specification.
   - Physical or chemical characteristics.
   - Material and methods of manufacture.
   - Performance.
4. By engineering drawing.
5. By miscellaneous methods.
   - Market grade.
   - Sample.
6. By combination of two or more methods.

Description by Brand
Two important questions arise in connection with the use of branded items. One relates to the desirability of using this type of description and the other to the problem of selecting the particular brand.

---

[2]Michiel R. Leenders and Harold E. Fearon, *Purchasing and Materials Management,* 10th ed. (Homewood, Illinois: Irwin, 1993), pp. 137-42.

Description by brand or trade name indicates a reliance on the integrity and the reputation of the supplier. It assumes that the supplier is anxious to preserve the goodwill attached to a trade name and is capable of doing so. Furthermore, when a given requirement is purchased by brand and is satisfactory in the use for which it was intended, the purchaser has every right to expect that any additional purchases bearing the same brand name will correspond exactly to the quality first obtained. Under certain circumstances, description by brand may not only be desirable but also necessary:

1.   When, either because the manufacturing process is secret or because the item is covered by a patent, specifications cannot be laid down.

2.   When specifications cannot be laid down with sufficient accuracy by the buyer because the supplier's manufacturing process calls for a high degree of that intangible labor quality called *skill*, which cannot be defined exactly.

3.   When the quantity bought is so small as to make the setting of specifications by the buyer too costly.

4.   When, because of the expense involved or for some similar reason, testing by the buyer is impractical.

5.   When users develop real, if unfounded, preferences in favor of certain branded items, a bias the purchaser may find almost impossible to overcome.

On the other hand, there are some definite objections to purchasing branded items, most of them related to cost. Although the price may often be in line with the prices charged by other suppliers for similarly branded items, the whole price level may be so high as to cause the buyer to seek unbranded substitutes or even, after analysis, to set its own specifications. Many articles on the market have, in spite of all the advertising, no brand discrimination at all. Thus, the purchaser may just as well prefer using trisodium phosphate over a branded cleaning compound costing 50 to 100 percent more.

A further argument, frequently encountered, against using brands is that undue dependence on brands tends to restrict the number of potential suppliers and deprives the buyer of the possible advantage of a lower price or even of improvements brought out by competitors through research and invention.

"Or Equal"

It is not unusual, particularly in the public sector, to see requests for quotations or bids that specify a brand or a manufacturer's model number followed by the words "or equal." In these circumstances the buyer tries to shift the responsibility for establishing equality or superiority to the bidder without having to go to the expense of having to develop detailed specifications.

Description by Specification

Specification, a particular form of description, is one of the best methods available for communicating needs to a supplier. Closely related are the concepts of standardization and simplification. Standardizing product specifications means defining uniform identifications for sizes, design, and quality, and gaining industrial acceptance for these standards. Simplification refers to a reduction in the number of product types and sizes that are accepted as standard.

Procedures for Developing and Reviewing Specifications

Specifications should be developed with input from sources both internal and external to the company. Internal input might come from a formal team, an informal team, or a coordinator. A formal specifications committee or team would include representatives from departments affected by the purchase. If there is no formal team or method in place to develop or review specifications, the supply manager must ensure that specifications, performance terms, or acceptance criteria are appropriately defined and documented before sending out a request for quotation. The supply manager will still need to involve representatives from affected departments or areas. A bid coordinator may work with the various departments within the organization to develop and document the bid package.

External input might come from suppliers and consultants, or industry standards, handbooks, and guides. Suppliers can be an excellent source of information on product specifications and should be consulted. However, suppliers should not be allowed to write the specifications. Consultants may be employed if no internal expertise is available. Industry standards, handbooks, guides, or other written sources may save time, effort, and inconvenience.

Advantages and Disadvantages of Specification
The advantages of specification include the following:

1. The process ensures that thought and study go into defining the need and the ways in which it may be filled.
2. A standard is set against which materials may be measured, thereby reducing delay and waste from improper materials.
3. Different sources may be asked to supply identical requirements.
4. The chance for equitable competition exists since suppliers are quoting for exactly the same thing. This is particularly critical to public purchasing.
5. Since performance terms may be specified, the supplier is responsible for performance.

However, there are limitations to the use of specifications, including the following:

1. It may be difficult or impossible to develop adequate specifications.
2. The immediate cost of using specifications may not be worthwhile in the long run if the requirement is purchased only in small quantities and does not conform to any definite standard.
3. If a specification is used, but the buyer asks the supplier to also quote on a standard item closely matching the specification, it may turn out that the standard item is appropriate.
4. The cost of testing to ensure that specifications are met may increase the immediate cost as compared to buying brand name.
5. A buyer may be lulled into a false sense of security by drawing up specifications when the specification is not really adequate.

6. Setting up elaborate specifications may discourage potential suppliers from placing bids.
7. Unless performance terms are included, the buyer bears the responsibility of adapting the item to its intended use provided that the item meets the description in the specification.
8. The minimum specifications set by the buyer are likely to be the maximum provided by the supplier.

Types of Specification

The three types of specification include:

- Specification by physical or chemical characteristics.
- Specification by material or method of manufacture.
- Specification by performance or function.

Specification by Physical or Chemical Characteristics
Specification by physical or chemical characteristics provides definitions of the precise material composition of the materials desired.

Specification by Material or Method of Manufacture
Usually, the more rigid the material specification, the greater the cost. Specification prescribing both material and method of manufacture is used in cases when special requirements exist. The buyer assumes most of the responsibility for results. These are also referred to as *design specifications*. A design specification provides a complete description of the desired item, including the composition of materials used as well as their size, shape, and mode of fabrication. By providing adequate design specifications, the purchasing manager avoids the possibility of having bidders perform duplicate work, encourages more competition by enabling firms who do not have the facilities for developmental work to participate in the bidding process, and enhances formal advertising as a means of purchase because bidders are able to bid on identical items.

Specification by Performance or Function
Specification by performance identifies the buyer's requirements in such terms as capacity, function, or operation. The buyer is more interested

in the service or function that the item will provide; the supplier decides details of design, fabrication, manufacture, and internal structure. The supplier assumes most of the risk for proper performance of the product.

Performance specification requires an understanding of the required function. Value analysis and value engineering focus on the definition of the basic function(s) because this is the area where the greatest value improvement is possible. Improper specification results when the function need is underestimated or overlooked. Buying quality will not solve the problem if the item is not the one really needed. By focusing on describing the basic function, usually in verb-noun combinations, like "transmit current," "cut material," or "contain liquid," the buyer can identify the function as clearly as possible. Secondary functions must also be identified, but it is the primary function that should be the main concern.

Description by Engineering Drawing

Description by a blueprint or dimension sheet is common and may be used in connection with some form of descriptive text. It is particularly applicable to the purchase of construction, electronic and electric assemblies, machined parts, forgings, castings, and stampings. It is an expensive method of description not only because of the cost of preparing the print or computer program itself but also because it is likely to describe an item that is quite special as far as the supplier is concerned and, hence, expensive to manufacture. However, it is probably the most accurate of all forms of description and is particularly adapted to purchasing items requiring a high degree of manufacturing perfection and close tolerances.

Miscellaneous Methods of Description[3]

Description by Market Grades

Purchase on the basis of market grades is confined to certain primary materials such as wheat, cotton, lumber, and cocoa. Purchase by grade is for some purposes entirely satisfactory. Its value depends on the

---

[3]Leenders and Fearon, pp. 143-44.

accuracy with which grading is done and the ability to ascertain the grade of the material by inspection.

The grading must be done by those in whose ability and honesty the purchaser has confidence. Even for wheat and cotton, grading may be entirely satisfactory to one class of buyer and not satisfactory to another class.

Description by Sample
Still another method of description is by submission of a sample of the item desired. Almost all purchasers use this method occasionally, but it is usually for a minor percentage of their purchases and then more or less because no other method is possible. Good examples are items requiring visual acceptance such as wood grain, color, appearance, and smell.

Combination of Descriptive Methods

An organization frequently combines two or more of the methods of description already discussed. The exact combination will depend, of course, on the type needed by the organization and the importance of quality in its purchases. There is no one best method applicable to any single product, nor is there for any particular organization a best method of procedure. The objective of all description is to secure the right quality at the best possible price.

Sources of Specification Data

Speaking broadly, specifications may be derived from three major sources: (1) individual standards set up by the buyer; (2) standards established by certain private agencies, such as other users, suppliers, or technical societies; and (3) governmental standards.

Individual Standards
Individual standards require extensive consultation among users, engineering, purchasing, quality control, suppliers, marketing, and, possibly, ultimate consumers. The task is likely to be arduous and expensive.

A common procedure is for the buying organization to formulate its own specifications on the basis of the foundation laid down by governmental or technical societies. To make doubly sure that no serious errors have been made, some organizations mail out copies of all tentative specifications, even in cases where changes are mere revisions of old forms, to several outstanding suppliers in the industry to get the advantage of their comments and suggestions before final adoption.

Standard Specifications

If an organization wishes to buy on a specification basis yet hesitates to undertake to originate its own, it may use one of the so-called standard specifications. These have been developed as a result of a great deal of experience and study by both governmental and nongovernmental agencies, and substantial effort has been expended in promoting them. They may be applied to raw or semimanufactured products, to component parts, or to the composition of material. The well-known SAE steels, for instance, are a series of alloy steels of specified composition and known properties, carefully defined and identified by individual numbers.

When they can be used, standard specifications have certain advantages. They are widely known, commonly recognized and readily available to every buyer. Furthermore, the standard should have somewhat lower costs of manufacture. Finally, because they have grown out of the wide experience of producers and users, they should be adaptable to the requirements of many purchasers.

Standard specifications have been developed by a number of nongovernmental engineering and technical groups. Among them may be mentioned the American Standards Association, the American Society for Testing Materials, the American Society of Mechanical Engineers, the American Institute of Electrical Engineers, the Society of Automotive Engineers, the American Institute of Mining and Metallurgical Engineers, the Underwriters Laboratories, the National Safety Council, the Canadian Engineering Standards Association, the American Institute of Scrap Iron and Steel, the National Electrical Manufacturers' Association, and many others.

While governmental agencies have cooperated closely with these organizations, they have also developed their own standards. The

National Bureau of Standards in the US Department of Commerce compiles commercial standards. The General Services Administration coordinates standards and federal specifications for nonmilitary items used by two or more services. The Defense Department issues military (MIL) specifications.

Standardization and Simplification

The terms standardization and simplification are often used to mean the same thing. Strictly speaking, they refer to two different ideas. *Standardization,* which means agreement on definite sizes, design, and quality is essentially a technical and engineering concept. *Simplification* refers to a reduction in the number of sizes and designs. It is a selective and commercial problem, an attempt to determine the most important sizes, for instance, of a product and to concentrate production or use on these wherever possible. For example, in most larger organizations, office furniture standards are established and options, if any, are limited, as are the suppliers. This not only permits a uniform interior design but also ensures fairness and reduces lead times and costs.

An electronics manufacturer bought 4,500 different capacitors from 40 different suppliers at a total cost of $40 million. A special commodity team composed of engineers, purchasing and quality control specialists, and production representatives from various divisions worked on this purchasing "niche" for about two years. The end result was standardization and simplification from 4,500 to 400 capacitors, a reduction in cost of about $5 million, a reduction in the number of suppliers from 40 to 6, and a reduction of up to six months in lead time, which improved new product introduction time significantly.

EARLY PURCHASING AND SUPPLIER INVOLVEMENT

Many of the tasks, systems, and procedures described in the beginning of this chapter relate to the operational side of the supply management function. Strategically, the question is, how can the supply function contribute effectively to organizational objectives and strategies? Because supply management brings in-depth knowledge and understanding of the external marketplace as well as of the organization's suppliers and competitors, it is critical that purchasing provide input to the corporate strategic plan just as its counterparts do in marketing or manufacturing.

Purchasing's contribution on the strategic side can take the form of analyzing the environment for short- and long-term changes in technology, capacity, and labor. The purchasing researcher should then be able to link environmental changes to products and services that may be affected and assess the probability of impact.

Both purchasing and the supplier can contribute through early involvement in developing specifications, particularly for new product development.

> Stryker Corporation, a medical products manufacturer, developed a defect-free product through the efforts of a team of Stryker engineers, purchasers, quality experts, and an electrical supplier. The result was an improvement in quality and customer service, a 10 to 12 percent cost reduction, and a 25 percent reduction in the development cycle.
>
> The goal of Stryker's purchasing department is to become one with key suppliers by making them feel they are an extension of Stryker's business. This ongoing process requires putting trust in a supplier. The focus is on building relationships and providing a highlevel of service and commitment to both internal and external customers.[4]

Specification-Related Requirements

Depending on the purchase in question, a variety of specification-related items should be included in the bid or quotation information sent to suppliers. Some of the items defined may be specific to the particular purchase. For example, engineered products are usually communicated with blueprints, material specifications, performance specifications (such as vibration, shock, temperature, corrosion resistance), and quality requirements. For capital equipment, specifications usually include statements on intended use of the equipment, the environment in which

---

[4]"High Performance Engines," *NAPM Insights*, January 1993, pp. 20-21.

the equipment will be required to function, the parts it will produce, capability studies, acceptance criteria, and sample performance reviews. For the purchase of services, specifications generally include detailed descriptions of the services expected, the time over which they must be performed, and an outline of the content of any final report. For purchases in general, the following items should be among those included in the Request for Quotation (RFQ), Request for Information (RFI), or Request for Proposal (RFP):

1.   Specific definition of what is being quoted, including quantity and timing (e.g., daily cleaning services for one year).

2.   Whether the supplier should quote alternate or substitute products or services.

3.   The quality performance specifications for the product or service (e.g., the quality standard that must be met, and documentation required either prior to or with the shipment).

Specification Problems and Abuses

Failure to define specifications can lead to buying goods and services that are not appropriate for the intended application. Incomplete specification can lead to delays or out-of-budget costs to modify goods and services to meet the needs of the organization. Additionally, three other problems are related to specifications:

- Over-specification resulting in higher costs.
- Slanted specifications written to a specific product or company, thereby eliminating competition.
- Use of custom specifications when generic specifications apply and would lead to a better price.

Quality has taken on new significance in the past decade. Customer satisfaction, Materials Requirements Planning (MRP), Just-In-Time (JIT), inventory reduction, time compression, reengineering, employee empowerment, and international competition have resulted in a complete rethinking of the role of quality. Initiatives such as Total Quality Management (TQM), Quality Function Deployment (QFD), zero defects,

Motorola's Six Sigma program, the Baldrige Award, the ISO 9000 Certification, Statistical Process Control (SPC), and a host of similar others have come into prominence. The simple fact is that customer satisfaction with respect to quality is to a large extent dependent on a supplier's quality performance. Jaguar determined in the early 1980s that over 60 percent of the quality problems in its cars stemmed from supplier quality failures. Therefore, suppliers must deliver the quality expected by the purchaser.

QUALITY

Quality can be defined in three ways:

- In absolute terms.
- Relative to a perceived need.
- As conformance with stated requirements.

In absolute terms, quality is a function of excellence, intrinsic value, or grades as determined over time by society generally, or by designated bodies in specialized fields. For example, gold is generally considered a high-quality metal. In business and industrial situations, quality is most accurately defined relative to the perceived need and/or conformance to stated requirements.

The simplest definition of quality is that the item or service meets specifications. However, the ultimate quality test is a use test. Does the item perform in actual use to the requirements of the final customer, regardless of conformance to specification? If the answer is no, the problem may lie with the original specification and not with the supplier, but the item will still be said to be of "bad quality." The idea that a misperceived need may result in inaccurate specifications that will result in purchased goods or services that do not meet the actual needs of the enduser or customer drives home the importance of need definition and specification development.

The concept of total quality incorporates a number of ideas. Total quality depends on the commitment and action of all people and organizations in the supply chain. Top management must drive the process, but every member of the organization is responsible for the achievement of total quality. Quality must be *designed* into products, processes, and services, rather than inspected in after the design. This requires the involvement of everyone in the supply chain. Customer needs must be thoroughly understood and communicated. Since inspection is decreased, quality assurance tools and techniques must be used consistently to prevent problems.

To produce quality goods or services, a supplier must meet a variety of specifications. Professor David Garvin of the Harvard Business School describes eight dimensions of quality:[5]

1. *Performance.* The primary function of the product or service.
2. *Features.* The bells and whistles.
3. *Reliability.* The probability of failure within a specified time.
4. *Durability.* The life expectancy.
5. *Conformance.* The meeting of specifications.
6. *Serviceability.* The maintainability and ease of fixing.
7. *Aesthetics.* The look, smell, feel, and sound.
8. *Perceived quality.* The image in the eyes of the customer.

Acceptance Testing

Determining whether a product meets specifications is accomplished through acceptance testing. The more frequently used methods of acceptance testing include the following:

Process capability testing. After a trial run at the supplier, the product is evaluated according to the characteristics that are affected by each specific manufacturing process.

Initial sample inspection. The product is examined and compared to a blueprint; a sample lot is tested for conformance to the requirements of the performance specification.

Incoming inspection. A sample of products from an incoming shipment is audited.

---

[5]David A. Garvin, "Quality on the Line," *Harvard Business Review* (September-October, 1983).

Supplier Certification

Many organizations are certifying their major suppliers so that shipments may go directly into use, inventory, or production. Certification involves evaluating the supplier's quality systems, approving the supplier's processes that are used to manufacture the products being purchased, and monitoring incoming product quality to ensure that it continues to meet requirements. The goal of certification is to reduce or eliminate incoming inspections.

Since the Uniform Commercial Code (UCC) does not make provisions for JIT arrangements under which a buyer does not inspect incoming goods, the buyer may be held responsible for all defects a reasonable inspection should have discovered, even if no inspection takes place. Therefore, it is important to document and have both parties sign the documentation for certified supplier arrangements. Routine audits performed on some predetermined frequency provide the monitoring function.

The advantages of supplier certification are increased product quality, reduced inspection costs, and the ability to work with suppliers on an ongoing basis to reduce process variation. Companies may designate suppliers according to certain categories based on the quality and reliability of their processes. The frequency and type of inspection would depend on the category or certification level of the supplier.

> At Burlington Northern Railroad, the top 200 of 5,000 suppliers will be audited as part of a supplier certification program. After a two-day audit, suppliers can be certified as a "Quality Vendor" based on their products and processes. To be classified as a "Preferred Vendor" the supplier must get at least 800 out of 1,000 in the areas of delivery, price, customer service, EDI, invoicing, packaging, and over, short, and damaged (OS&D). The designation is good for two years.[6]

---

[6]Landon J. Napoleon, "Supplier Certification: Where Are We Headed?" *NAPM Insights*, November 1991, p. 10.

Inspection and Testing

Inspection and testing may also be performed to ensure quality and minimize problems. Testing may be done before committing to a supplier to ensure that the supplier can meet the quality standards of the specification. A buyer may also test to compare the capabilities of several different suppliers. Inspection may be necessary upon receipt of items from the supplier. The purpose of inspection is to ensure that the delivered item meets specifications.

Specifications should include the type and frequency of inspection so that there are no problems agreeing on the results of an inspection. Inspection procedures should be established by both buyer and supplier in light of the cost of inspection relative to the benefits. The two basic types of quality inspection are 100 percent inspection and sampling.

Use tests or engineering tests may be performed either in the buyer's facilities or in commercial laboratories. Samples should be tested in well-designed experiments so that the suppliers know their samples received a fair evaluation. A use test allows the buyer to evaluate the item in the particular purpose for which it was designed. Components often cannot be placed in actual use prior to completion of the finished product. In these instances, customers can provide information on the performance of components from various suppliers.

Inspection is expensive and it does not add value if the quality is acceptable. Ideally, inspection would be unnecessary because the buyer and supplier have built quality into the design and production and the supplier provides reliable supporting evidence of outstanding quality performance. The reality is that many organizations have not reached this goal, and inspection is required. The type, thoroughness, and frequency of inspection depend on the quality assurance program developed cooperatively by the buyer and the supplier.

One hundred percent inspection may not achieve the goal of 100 percent quality because the inspection process may also be subject to error such as worker boredom or fatigue or inadequate facilities or methods. Some tests are too destructive or cost-prohibitive for 100 percent inspection.

Sampling

To avoid the problems associated with 100 percent inspection, a sample of output may be inspected and tested. The key is selecting a sample that is representative of the total output or population. Random sampling is commonly used. *Random sampling* means that each unit of output has an equal chance of being selected for inspection and that it is reasonable to assume that the results of the inspection for that sample can be applied to the total output or population. If all output share the same characteristics, then any group of units may be selected and tested. If output have different characteristics, then consecutive numbers may be assigned to the output and random number tables or computer programs may be used to select the items for inspection.

Quality Assurance Concepts

Process Quality Control

In supplier selection and ongoing supplier performance evaluation, the buyer is concerned with ensuring that the supplier is able to meet the quality specifications. One important tool is measuring the capability of the process(es) used by the supplier and comparing that capability to the quality specifications set by the buyer. As long as the buyer's quality specifications fall within the process capability of the supplier, the buyer should be reasonably certain of receiving satisfactory items. If, however, the buyer specifies a narrower quality range than the supplier's process is capable of achieving, the buyer will probably have to worry more about inspection, rework, and scrap.

The first step in quality assurance is making sure that the supplier's process capability and the buyer's quality range mesh. If the natural range of the supplier's process is wider than the range of the buyer's quality requirements, the buyer must negotiate with the supplier to have the supplier narrow the natural range through process improvements such as operator training or machine improvements. If it is not economically feasible or the supplier is unable or unwilling to make improvements for some reason, then the buyer may seek another supplier rather than incur the extra cost of inspection, rework, and scrap.

Process capability refers to the ability of the process to meet specifications consistently. Since no process can produce the same exact results each time the activity is performed, it is important to establish what kind of variability is occurring and eliminate as much variability as possible. A process capability study identifies two types of variability: (1) the common causes or random variability and (2) special causes.

Common causes of variability may be related to machines, people, or manufacturing. For instance, machine lubrication, tool wear, or operator technique would be common causes of inconsistent output. These causes are often a natural part of the system and do not fluctuate. The only way to reduce common causes is to change the process. Taking common causes into consideration, the output distribution about the mean quality level results in the natural capability for a particular process.

Special causes or outside, nonrandom problems such as wear and tear on machinery, material variation, or human error must also be identified and eliminated; otherwise the output will fall outside the acceptable quality range. Statistical process control procedures are primarily concerned with detecting and eliminating special causes.

In determining whether a process is stable, the supplier must determine the natural capability of the process, and whether the upper and lower specification limits meet the specifications of the buyer. When a process is "under control," the supplier can predict the future distributions about the mean. For a process to be capable and in control, all the special causes of variation in output have been eliminated and the variation from common causes has been reduced to a level that falls within the acceptable quality range specified by the buyer.

A process is capable when the process is in control, stable, and predictable and when the process averages a set number of standard deviations within the specifications.

Cpk

The Cpk statistic is used to measure and describe machine or process capability relative to specification requirements by determining the percentage of parts that a process can produce within design specifications. Cpk is defined as the *lower* of either of the following:

$$\frac{\text{Upper Specification Limit - (Process Average)}}{\text{Process Spread}}$$

or

$$\frac{\text{Lower Specification Limit - (Process Average)}}{\text{Process Spread}}$$

Process spread is equal to three standard deviations of the output values, or the spread on one side of the process average. The lower the Cpk, the more capable the process is of producing parts that are consistently within specification.

Statistical Process Control (SPC)

At times the process may be "out of control," which means that statistically the distribution about the mean can no longer be predicted. The result is output that does not conform to the quality specifications. This unpredictability is caused by special causes or nonrandom events such as a problem with material quality, excessive human error, or worn tools.

Statistical process control is a technique for detecting nonrandom changes in the process. Once detected, the cause of the problem must be determined and the problem corrected. Often the machine operator is able to detect, identify, and correct the problem before many poor-quality items are produced.

The key tool in the detection of special causes is the *quality control chart*. First, upper and lower control limits are set to establish the desirable normal operating range. Quality control charts then track the means of the output (X bar chart) and the dispersion about the mean (R bar chart). By tracking random samples of the total output the operator is able to detect if the machine process is in or out of control. Control charts must be tracked continuously to get a picture of the trend rather than a snapshot in time.

Statistical process control works in repetitive process situations such as continuous or semicontinuous manufacturing or other standardized procedures. Determining the normal operating range of a process and whether the process is stable exposes problems caused by external events. The cost of implementing SPC must also be compared to the length of time the process will be in operation and the expected benefits of SPC.

ISO 9000

ISO 9000 is a series of five standards developed in the 1980s by the International Organization for Standardization to document, implement, and demonstrate quality assurance systems. ISO 9000 requirements do not refer to products or services, but to the systems that produce them. Through a combination of external and internal audits a company can ensure that a quality system is in place to recognize their own level of quality and customer satisfaction.

The five levels of approval within ISO 9000 are:

ISO 9000   Provides basic quality assurance definitions and guidelines for selection and use of other standards.

ISO 9001   Deals with quality systems for design, development, production, installation and customer service.

ISO 9002   Provides a quality assurance model for conformance in production and installation, generally applicable to companies doing little product development.

ISO 9003   Provides a model for conformance in final test and inspection standards.

ISO 9004   Provides guidelines for the establishment and documentation of a total quality system. It assists in developing and identifying the relationships between inputs necessary to assure quality.

QUANTITY

Specifications tell *what* to buy, but not *how much* to buy. The quantity decision is complicated by several factors. How much is needed right now, next month, in six months? What will be the cost if inventories are held or if material is bought as it is needed? Do required quantities lead to price breaks or price premiums? Are shortages anticipated? If so, what buying plan is most advantageous from a price and supply assurance standpoint? The list of questions to determine how much to acquire may seem formidable.

Decisions often have to be made far in advance of actual usage; hence forecasts are critical. Costs may be driven up if material is held in inventory; or if materials run out, a premium may be charged for certain quantities or delivery terms. These and other quantity concerns are covered in Book 3 of this series.

FORECASTING

If the quantity decision is pivotal in achieving organizational objectives in a cost-effective manner, then forecasting is one critical tool linking organizational objectives and strategies with supply objectives and strategies. The role played by the supply manager in forecasting varies among organizations. At the very least the buyer must be aware of forecasts and adjustments and inform the supplier in a timely manner without harming buyer-supplier relations.

In keeping with the supply chain concept, forecasts are derived from knowledge of customer needs and wants and then worked through the entire organization, including sales, purchasing, and logistics. Forecasts are essential but are unreliable. Who makes the forecast and how it is derived and monitored has a tremendous impact on the entire organization. Whether or not purchasing is involved in the forecasts, the results of a poor forecast, either over or under actual needs, will be remembered as over- or understocking. Purchasing has a lot to gain or lose in terms of reputation, reliability, and professionalism if the forecast is substantially off.

The buyer or supply manager must assess the forecast carefully and devise a procurement plan that takes into consideration the probability of all possible outcomes. If the range of demand is wide, the purchaser needs to keep the supplier updated regularly. A supplier may believe that the purchaser was merely trying to gain a favorable price if demand falls below forecasts. On the other hand, if demand greatly exceeds the forecast, the supplier may incur additional costs in overtime, rush orders, and altered production schedules. In either situation, the buyer must remain aware of the changes and keep the supplier informed.

The two basic types of forecasts are qualitative and quantitative. Qualitative forecasts rely on the collective opinions of experts such as sales staff and district sales managers. Quantitative forecasts use past data to predict the future. Quantitative models may rely on leading indicators that are believed to cause changes in sales or on repetitive sales patterns over time. The latter, known as time series forecasts, consider the six aspects of constant value, trend, seasonal variations, other cyclical variations, random variations, and turning points, to identify the pattern and make a forecast.

Forecasting techniques are covered in greater detail in Book 4 of this series.

TIMING ISSUES

Obviously, timing issues abound in purchasing. First, when does the requisitioning department, unit, or person need delivery? Second, how long will it take for the procurement process? Third, what is the supplier's lead time? Fourth, what are the benefits or penalties for early or late delivery?

When Is Delivery Required?

Determining a delivery date is not quite as simple as it appears. The requisitioner may not be certain of a required delivery date. Moreover, even after delivery, it may take time for receiving, inspection, and materials control and handling before the requisitioner can actually use the requested items. In anticipation of potential delays, the requisitioner may add a safety margin onto the requested date. The end customer may change delivery requests. Some requisitioners automatically request "as soon as possible (ASAP)" and create unnecessary rush orders.

Rush Orders[7]

Rush orders cannot always be avoided; emergencies do arise to justify their use. Sudden changes in style or design and unexpected changes in market conditions may upset a carefully planned material schedule. Breakdowns are inevitable, with an accompanying demand for parts or materials not normally carried in stock.

However, some rush orders cannot be justified on any basis. Such requisitions arise because of (1) faulty inventory control, (2) poor production planning or budgeting, (3) a lack of confidence in the ability of the purchasing department to get material by the proper time, and (4) the sheer habit of marking the requests "rush." Whatever the cause, such orders are costly. This higher cost is due in part to the greater

---

[7]Leenders and Fearon, pp. 71-73.

chance of error when the work is done under pressure. Rush orders also place an added burden on the seller, and this burden must directly or indirectly find its way into the price paid by the buyer.

What can be done to reduce the seriousness of the problem? For an excessive number of rush orders that are not actually emergency orders, the solution is a matter of education in the purchasing procedure. In one company, for example, when a rush requisition is sent to the purchasing department, the department issuing such an order has to secure approval from the general manager after explaining the reason for the emergency requirement. Furthermore, if the requisition is approved, the extra costs, so far as they can be determined, are charged to the department ordering the material. The result has been a marked reduction in the number of such orders.

In organizations using MRP, delivery has to be as scheduled and buyers often have the job of buyer-planner. In a JIT mode, delivery may be triggered by a kanban and may be required in a matter of hours or even minutes.

The Length of the Procurement Process

The procurement process itself—determination and specification of need, supplier search and evaluation, determination of appropriate contract format, request for information (RFI), request for bid (RFB), or request for quotation (RFQ), analysis of bids, and placement of the order—may take a considerable amount of time. In a utility, for example, the search for an appropriate engineering consulting firm for a major capital project took about six months.

It is not surprising that in organizations and situations where speed is of the essence, the acquisition process itself faces major pressures for time compression. Early supplier involvement (ESI), particularly important in the design of new products or processes, is an attempt to reduce acquisition time and also design, testing, and manufacturing run-in-time. In reverse marketing, the acquisition process time is separated into three categories:

> *Desperation time*—not enough time to do a good job of procurement. Prime emphasis is on minimizing the damage.

Normal supply time—sufficient time is available to do a sound job of procurement and to reap the benefits of internal synergy.

Reverse marketing time—enough time to help create or change a supplier to get the benefits of external synergy.

Supplier Lead Time

Suppliers need time to satisfy the purchaser's order. When a supplier has finished goods inventory, lead time may represent order processing, picking, packaging, and delivery time. When a custom requirement is placed, the supplier's lead time may include design, tooling development, manufacture, testing, and delivery time. For major products such as turbines for power generation or the trains for the English Channel project, lead time may be up to five years or more.

In periods of shortages lead times tend to lengthen, while in periods of overages lead times shorten. The purchaser needs to be aware of requisitioner needs and internal and external processing times. The purchaser should also assist the requisitioner with the available information to avoid unpleasant surprises.

Rewards or Penalties for Early or Late Delivery

It used to be assumed that early delivery was superior to on-time delivery and late delivery inevitably bad. The JIT philosophy shows that early delivery is undesirable because space may not be available and the system cannot handle it. Extra handling costs may also be incurred. Anyone familiar with project management will recognize that early delivery for items on the critical path may be beneficial, whereas early delivery of noncritical requirements is, at best, immaterial or, at worst, dysfunctional. Early delivery for a time-critical requirement on a new product development may allow the organization to achieve earlier market entry and the associated benefits.

Late delivery—late in the sense that it did not conform to the original date on the purchase order, may or may not have serious consequences. If it has no impact on customer satisfaction, operational costs, or competitiveness, a late delivery may be acceptable. On the other hand, with JIT customers or internal users dependent on on-time delivery, late delivery may have serious adverse consequences.

The alert buyer will be aware of the various benefits and trade-offs involved in early and late delivery and will ensure that maximum benefits are pursued and dire consequences avoided.

CONCLUSION

The first step in the acquisition process is the determination of organizational needs. This chapter has focused on classification, description, quality, quantity, and timing. Next, we will discuss where to find suppliers who might be capable of satisfying these needs.

KEY POINTS

1. Determination of organizational needs is the first step in the acquisition process.

2. Classification of needs by type of requirement—by ABC category, and by strategic or operational, repetitive or nonrepetitive categories—permits the focusing of purchasing effort where it will contribute maximum value.

3. A need can be described in many ways.

4. Complete and accurate description is essential to ensure that the internal need will be satisfactorily met and that the supplier will be able to meet the requirement.

5. The purchase requisition (PR) is the common method used by a requesting unit, department, or person to communicate a need to the purchasing department.

6. Requisitions and descriptions need to be reviewed carefully by purchasing to ensure appropriateness, proper authorization, and conformance to laws, policies, and procedures.

7. In many public organizations purchase requisition review also requires a check against the requisitioner's budget to ensure that sufficient funds are available for the purchase.

8. A large percentage of the cost of a requirement is determined at the specification stage.

9. Standardization and simplification offer substantial opportunities for value improvement.

10. Early purchasing and supplier involvement, especially at the design and specification stages, will ensure that requirements will be met on a timely basis, will represent good value, and will be procurable as well as manufacturable.

11. Quality includes performance, features, reliability, durability, conformance, serviceability, aesthetics, and perceived quality.

12. Specifications should include the testing procedures to be used to assure conformance.

13. Supplier certification aims to reduce duplicate inspection.

14. Buying the right quantities may be complicated by inaccurate forecasts of requirements.

15. The timing of deliveries needs to take into account internal time requirements as well as the lead time required by the supplier.

CHAPTER 3

POTENTIAL SOURCES OF SUPPLY

Most purchasers agree that selecting a supplier is the most important decision to which they contribute. Choosing the right supplier will ensure that the purchaser's operational and strategic needs will be satisfied. To make the right selection, the purchaser must first identify potential sources of supply and analyze and evaluate the sources based on predetermined criteria. Often the purchaser will develop approved lists of suppliers.

COMPETITION AND COOPERATION

By force or design, the purchaser may encourage, support, or develop competition or cooperation among suppliers. Traditionally, purchasers used several suppliers to maintain competition among them and to hedge the risk of one supplier's failing to meet the business needs of the purchaser. Recent trends of closer purchaser-supplier relationships have led to efforts to create cooperation with a few key suppliers. Balancing the desire for competition with the benefits of cooperation is a continuing challenge for purchasers.

The degree of competition depends on the external environment and on the needs of the purchaser. How the purchaser proceeds with the analysis and selection process depends partially on the degree of competition. There are several types of competition:

- *Full and open competition* exists when there are many buyers and sellers and there is no collusion among them over the long term. Product improvements are promoted and price containment is possible in this setting.

- *Limited competition* exists when there are few sellers. This can occur nationally if demand is great and locally if the few sellers follow each other's behaviors and pricing. Purchasers have fewer choices of supply, and pricing is more fixed or firm.

- *Technical competition* may not exist when only one or a few suppliers provide the technology or hold the patents on a special product.

- *Sole sourcing* exists when there is only one source available. Local telephone and water companies are examples. A sole source may make a special product or technology that no one else does.

- *Single sourcing* exists when an organization decides to select one supplier although many are available. Partnerships, strategic alliances, and just-in-time relationships are often examples of single sourcing. Some of the reasons for single sourcing are consistent quality, administrative simplification, joint schedule planning, smooth logistics links, and joint costing and pricing benefits.

STRATEGIC SOURCING CONSIDERATIONS

A number of sourcing choices are open to the buyer; each has its own set of advantages and disadvantages. The buyer must choose the type of supplier, the size of the supplier, the number of suppliers for each requirement, and the location of the supplier. These choices are discussed in detail.

Manufacturers versus Resellers

When buying in bulk, it is often preferable to buy from manufacturers because the cost will not include a distributor's markup or profit. Also, special design or technical support may be available from the manufacturer.

Resellers, including distributors, wholesalers, and retailers, may be chosen if the buyer has limited warehouse facilities, needs only a small quantity, or needs to purchase small amounts of a wide range of products by many different suppliers. The decision to use a distributor is often based largely on the need for services. Using distributors or wholesalers may cost less than buying directly from the manufacturer. Marketing costs are spread over a variety of items and transportation costs may be less for distributors because they can buy in car- or truckload lots.

The buyer may wish to avoid the distributor and buy directly from the manufacturer if no selling effort or service is rendered by the supply house. Buyers will sometimes try to buy from small manufacturers that do not have a widespread distribution network, or they may secure a special service such as dedicated staff from a distributor.

Systems contracting and stockless buying systems combine small requirements and place a large variety of requirements such as electrical supplies with one distributor.

Size of Supplier

Another consideration in supplier selection is the size of the supplier. Should a buyer favor one size of supplier over another? Several factors affect this consideration, including the size of the purchasing organization, the nature and volume of the requirement, the need for proximity to the supplier, and the after-sales service requirements.

Generally, a large supplier is sought when the volume required is large and quality, price, and total cost of ownership are critical. Small, local suppliers are usually selected when flexibility, speed of response, and availability are more important than price. Medium suppliers fall somewhere in between.

There are many exceptions to these generalizations, especially with the current focus on customer service, which has led many large organizations to develop better service capabilities. The focus on cost reduction may place smaller firms at a disadvantage because larger firms may have greater stability and greater resources. Social responsibility and government pressure for minority supplier development are two factors increasing the likelihood of a large buyer-small supplier matchup.

Number of Suppliers

The purchaser must also decide if conditions warrant using a single or sole supplier or multiple suppliers. Deciding what percentage of the business to give to a supplier depends on the level of risk the purchaser and the buying organization are able and willing to take.

The reasons for choosing one source include:

1. Superior quality.
2. The order is too small to split.
3. A close buyer-supplier relationship, such as a partnership, is preferred.
4. An ongoing long-term contract exists.
5. A supplier owns a patent or process.
6. Price or freight discounts are available.
7. Greater supplier commitment.
8. The cost to duplicate equipment or setup is high.
9. Ease of scheduling.
10. JIT capability.
11. Stockless buying or systems contracting is desired.
12. Time and resources available to develop more effective buyer-supplier relationships.
13. The Japanese have been successful using single sources.

The reasons for multiple sourcing include:

1. It is common practice.
2. Suppliers may give better prices and service with the pressure of competition.
3. Supply assurance if disaster strikes.
4. To avoid becoming the sole support of a supplier.
5. Flexibility—unused capacity is often available to meet current and future needs.
6. Backup from several suppliers.
7. Strategic reasons such as supply security and military preparedness.

8.   Government regulations may require multiple sources.
9.   Potential and new suppliers can be tested while other suppliers provide the bulk of the requirements.

Location of Suppliers

The decision to use a local, national, or international source depends on a number of issues. Sometimes a purchaser may be forced to use a particular supplier, perhaps a subsidiary of the same firm. Or the purchaser may act on government regulations or social responsibility and purchase from small, disadvantaged, or labor surplus area suppliers. In general, most purchasers prefer local sources. This preference is complicated by the purchaser's basic responsibility to buy best value; the search for the best buy may lead to a source in another city, state, province, or country. Also, technological advancements in communication and the globalization of the economy cloud the definition of local. Is a local source in the same city, county, state or province, or country?

A local source may be preferred for economic reasons and out of a sense of social responsibility. A local source may also provide more dependable service and have greater flexibility in meeting the purchaser's requirements. Local sources may provide faster delivery at lower costs, will usually be more willing to tailor their business to the purchaser's needs, and may maintain a higher level of service. If financial strength, facilities, and capabilities are equal for a local and a distant supplier, the purchaser may well prefer the local source because of these advantages.

Social responsibility also leads purchasers to prefer local sources. A company owes much to the local community that provides employees, financial support, and schools, churches, and housing for employees. It is good public relations to work with local businesses and generate goodwill toward the organization.

Some companies support and develop local suppliers to improve both the company and the local economy. During the "gold boom" of the mid-1980s, mining companies in Elko County, Nevada, encouraged suppliers to open local branches because of the difficulties created by distant suppliers. Problems with shipping, long lead times, high freight budgets, increased inventory cost, and the potential for extended downtime led the mining companies to support the development of the local economy.

Newmont Gold Company in Carlin, Nevada, purchases over $100 million of goods and services to support its mining activities. In 1992, 57 percent of these purchases were from local suppliers in Elko County. At the time, these local suppliers accounted for only 14 percent of Newmont's total supplier network.[1]

Purchasing Internationally

Buying from national or international sources increases competition and may provide better quality, lower prices, and a wider range of goods and services. Domestic sources may be preferred over international ones because the same laws, transportation links, and communications apply for the purchaser and supplier. Some of the special considerations when sourcing internationally are exchange rates, payment processes, duties, transportation costs, timing, and laws.

Exchange Rates
The rate of exchange refers to the price at which one currency can be bought with another currency or gold. The relative position of the exchange rate between two nations will affect the price level of traded purchases and sales. Currency levels are often unpredictable and therefore constitute a true risk factor in international business.

International Payments
Payments in international transactions are usually in the form of letters of credit or bills of exchange. The *letter of credit*, used most frequently, is a specific line of credit granted by the importer's bank. A *bill of exchange* is a document issued by the seller instructing the purchaser to make payment in full at a specified point in time, similar to a check that instructs a bank to pay on a depositor's behalf. If and when the buying

---

[1]"The Local Supplier Base," *NAPM Insights*, February 1993, p. 13.

firm signs the document, it becomes a negotiable instrument and is a trade acceptance. Generally, international purchasing ties up capital longer than does domestic purchasing.

Duties and Taxes

Duties are taxes levied by the government on the importation, exportation, or use of goods. Most goods entering the US market have duties assessed on them. The North American Free Trade Agreement eliminated or altered duties on many goods and services between the United States, Canada, and Mexico. The three major types of duties are specific, ad valorem, and compound. *Specific duties* are charged as a specified rate per unit, for example $15 for each crate. *Ad valorem duties,* the most prevalent, are charged as a percentage of the appraised value (e.g., 3 percent ad valorem). *Compound duties* combine specific and ad valorem rates, for example $2.15 per gallon plus 8 percent ad valorem.

International Transport

The benefits of international procurement must be weighed against the costs due to longer transportation links. In North America, shipments may be made via truck, rail, air, or ship. Overseas shipments may be made by air transport to reduce transit time, but the higher costs may outweigh the benefits of speed. Shipping services provide cheaper international transportation, but at much slower speeds. Water transportation is typically used for large-volume purchases and the transport of raw materials. Three basic forms of ship service are: (1) Liners which carry freight for the general public with scheduled fixed routes; (2) Tramp ships which move shipload cargoes wherever they might originate or terminate, and (3) Private ships which are owned by the firm moving the goods; these are typically used in the banana, automobile, and lumber industries.

The laws of the countries of both purchaser and seller will affect the transaction and may complicate the flow of goods. Import licenses, custom duties, and quotas may restrict imports. Complex documentation required by governments may include export licenses, import declarations, certificates of origin, commercial invoices, customs invoices, insurance policies, and bills of lading. As Mike Unstead-Joss, Du Pont's quality systems manager for European manufacturing says, "When I'm sitting here in Belgium evaluating suppliers, I do not distinguish between firms in Germany, Texas, or Brussels as long as they are registered to ISO 9000. There's no Fortress Europe built around ISO 9000. It's an international standard."[2]

TYPES OF SUPPLIER RELATIONSHIPS

Depending on the nature of the purchaser's operational and strategic needs and market opportunities, the purchaser has a choice not only of supplier but also the type of relationship to foster with the potential supplier.

Traditional Suppliers

The traditional supplier-purchaser relationship is often characterized by frequent supplier switching, emphasis on price over other considerations, adversarial relationships, and short-term contracts.

Preferred Suppliers

A preferred supplier provides the desirable quality, delivery, or price levels and reacts positively to unforeseen needs such as changes in volume, specifications, and service problems. A preferred supplier takes the initiative in suggesting better ways to serve the customer, finds ways to enhance the products, and warns the customer in advance of factors that may affect the purchaser's operations.

Certified Suppliers

A certified supplier is one whose quality control system is integrated with the purchaser's system, thus establishing a larger quality assurance system. Total costs of quality assurance are reduced through the elimination of duplicate efforts in inspection and other quality control activities.

---

[2]Rita Grenville, "ISO 9000 Outside US Borders: Status and Trends," *NAPM Insights*, November 1992, p. 26.

Prequalified Suppliers

Firms on the purchaser's list of approved suppliers and bidders are prequalified to do business with the purchaser. Prequalification requires an in-depth analysis of the supplier's financial strength, facilities, location, size, technology, labor status, management, costs, terms, references, and other factors.

Supplier Partners and Strategic Alliances

Suppliers that have a close relationship with a buying firm are often referred to as partners. Partnerships are based on mutual trust and support, information sharing, and joint work on continuous improvement. When an unusually close supplier relationship is created for strategic reasons, it is often called a strategic alliance. Partnerships and strategic alliances are marked by long-term arrangements, large volume commitments, and joint product development and production planning efforts. Partnerships and strategic alliances are discussed extensively in Book 4 in this series.

> A strategic partnership allows both parties to leverage their strength and pool their resources to create competitive advantage. Dow Chemical Company set up an agreement with Domtar, Inc., of Montreal to recycle plastics used in soft drink bottles and cartons. Dow supplies the technology to utilize the plastic and Domtar, which already recycles paper, supplies the plastic. This partnership results in new product development, quality improvement, problem solving, and crisis intervention.[3]

---

[3]Lt. Jane A. Barclift, US Navy. "Strategic Partnership: Competitive Advantage or Risky Business?" *NAPM Insights*, January 1991, p. 7.

EXISTING VERSUS NEW SUPPLIERS

There are advantages and disadvantages to continuing with an existing supplier or to choosing a new source.

Market Conditions

Market conditions can influence the decision to use an existing supplier or seek a new supplier. If working with a single source, the purchaser must monitor market conditions to ensure that the supplier remains competitive. One of the greatest fears of the purchaser in a single source situation is that the supplier will include unreasonable price increases without the purchaser realizing it. By seeking new sources, the purchaser can guard against such increases. If, however, shortages loom on the horizon, the purchaser may benefit from remaining with an existing supplier and even guard against the impending shortage by locking into a longer-term arrangement with the supplier. Applying accurately read market conditions to the current and future needs of the organization allows the purchaser to contribute to the strategic goals and objectives of the organization.

Adequacy of Competition

New suppliers may represent expanded supply options, remove the purchaser from a single source situation, offer additional technical services, or more effectively meet other unmet needs.

Purchasers should, however, beware of switching to a new supplier hurriedly because new suppliers sometimes bid low to attract business and then raise prices based on "new costs." Avoiding an over dependence on one supplier can help avoid this tactic. Familiarity with the cost elements of the good or service will also help the purchaser avoid switching suppliers when only short-term benefits are likely.

Long-Term Needs and Long-Term Relationships

Meeting the long-term needs of the organization should be the driving force behind the strategic decisions made by the purchaser. However, meeting current needs is also important, and the failure to meet current needs is immediately noticed by the purchaser's internal customers. Therefore, the purchaser must balance these sometimes competing needs in source selection decisions.

Factors favoring long-term relationships are prior commitments, a successful past relationship where standards of quality, service, delivery and price are met, an ongoing long-term contract with a preferred supplier, or the wish to build supplier goodwill by granting more business to reward good service. Uniformity and continuity of supply can result from long-term relationships. Partnering arrangements are entered into on the basis of these types of arguments.

A counterargument can be made for seeking new sources of supply as a means of obtaining long-term needs. By seeking new sources of supply, the purchaser may protect the company's supply lines in case a supplier is unable to perform as promised.

Product Complexity and Technology Changes

If a supplier is the sole owner of a certain patent or process, this may dictate the purchaser's source selection. Likewise, if the product involves costly tool, die, mold, or setup charges, the expense of duplicating this equipment will also discourage the use of new suppliers. However, staying with an existing supplier may cause the purchaser to miss out on new technology offered by another source.

Urgency of Need

The amount of time available to the purchaser also influences the supplier selection process. Since switching to a new supplier requires time and expense to get the supplier up and running, it may not be practical to use a new source.

Quality Expectations

The quality demands of the purchaser may already be met by an existing supplier, leaving little motivation to seek other sources. However, if supplier quality is marginal or if problems are recurring, the purchaser may seek a new supplier that can meet the quality standards or is more able and willing to provide superior service.

Need for Modification of the Supplier Base

The supplier base should be evaluated and changed to reflect the best possible list of sources to meet current and future needs. The development of new sources will help keep present suppliers alert to the expectations of the buying organization and aware of the pressures of competition. Purchasers should give suppliers ample opportunity to correct any problems that arise, beginning with keeping the lines of communication open between the two organizations.

Changes in the Supplier's Organization

Major changes in a supplier, such as retirement of key personnel, a merger, or a takeover, may cause a disruption in supply or quality, forcing the purchaser to seek new sources. Analysis of the supplier's organization and management should help prepare the purchaser for possible disruptions and may lead to the decision to split business among several suppliers.

Reverse Marketing or Supplier Development

In some instances a suitable source may not already exist, and the purchaser may need to develop or create a source to meet either a current or a future requirement. New technology is the normal reason why a new product or service needs to be developed. Strategic, political, economic, or environmental causes may also create the need for new sources.

Reverse marketing or supplier development reverses the traditional purchaser-supplier relationship in that the purchaser aggressively and selectively prepares proposals to a few key suppliers to maximize their strategic contribution. The benefits to both purchaser and supplier may be limited to the particular order, or they may be more far-reaching, such as financial, technical, management processes, skills, quality levels, reduction of marketing effort, and smoother manufacturing levels.

One of the most compelling reasons to engage in reverse marketing or supplier development is the contribution supply management or purchasing can make toward achieving the organization's long-term goals and objectives. Purchasers are required to be more aggressive with suppliers to meet the future needs of the organization. Three external forces—technology, international trade, and management focus on quality and continuous improvement—push the purchaser to create suppliers instead of just selecting them.

SOCIAL, POLITICAL, AND ENVIRONMENTAL CONSIDERATIONS

Besides economic considerations, there are social, political, and environmental issues to evaluate when selecting sources.

Socially and Economically Disadvantaged Suppliers

Government legislation, social responsiveness, and the desire to develop alternate sources of supply have led to the development of minority purchasing programs, sometimes called socially and economically disadvantaged supplier programs. Minority businesses are generally defined as those in which at least 51 percent of the ownership is held by members of a minority group.

In 1968 the Small Business Administration established a program to channel federal purchases to socially or economically disadvantaged small businesses. In 1969, the US Office of Minority Business Enterprise was established within the Department of Commerce to mobilize federal resources to aid minority-owned businesses. In 1978, Public Law 95-507 mandated that bidders for federal contracts in excess of $500,000 for goods and services and $1 million for construction submit, prior to contract award, a plan including the percentage goals for using minority businesses. In 1983, President Reagan signed Executive Order 12432 directing all federal agencies to develop specific goal-oriented plans for expanding procurement opportunities to minority businesses.

Typically, organizations committed to buying from minority- or women-owned businesses include in the policy manual a statement signed by the CEO or some other top executive; develop a separate procedure for dealing with these suppliers; and designate a minority supplier coordinator. The benefits of establishing an active program are enhancement of the company's image in the community and with minority and women customers. The impediments to programs for small and disadvantaged businesses and women-owned businesses are active and passive resistance and hostility that may be encountered within the organization and lack of purchaser training in minority business matters. Education and training can help overcome these barriers. The minority supplier coordinator can provide guidance in most cases. Assessing the program requires accurately coding suppliers as small, large, and minority- or women-owned in the supply base and periodically reporting on all supplier activity within each category.

> Pepsi-Cola Company developed the HUB (Historically Underutilized Businesses) process in 1982. Cumulative spending in 1990 exceeded $410 million. One of the successes of the HUB process led to two former truck drivers of a majority sweetener hauling firm becoming entrepreneurs without putting up any capital. Pepsi-Cola's director of sweeteners asked the majority firm to identify two of its best drivers—one African American and one Hispanic—and to lease two $50,000 stainless steel tankers to them to haul sweetener to Pepsi-Cola. Two years later, the two men, Jerry Woodward of Woody's Trucking and John Aguilar of Vernon Transportation, hired six to eight drivers and either leased or owned their own equipment.[4]

Environmental Considerations

Environmental considerations in the acquisition process have received greater attention in recent years. Federal regulations for the use,

---

[4]Betty L. Darrell, "Adding to a Solid Foundation," *NAPM Insights*, October 1991, pp. 2-23.

transportation, and disposal of hazardous materials significantly affect how the purchaser performs his or her job, as do workplace right-to-know laws and emergency preparedness requirements.

The main questions for the purchaser to answer are: (1) should the organization purchase materials that may increase environmental concerns? and (2) should we purchase from suppliers that we know are not following sound environmental practices? The impact of environmental issues is felt at every stage of the acquisition process, from need definition to disposal. The purchaser must determine and give the appropriate weight to environmental concerns in every purchase.

Environmental issues are discussed in greater detail in Book 4 of this series.

LOCATING POTENTIAL SOURCES OF SUPPLY

There are a number of resources for locating potential sources of supply that are not currently suppliers. For existing organizations, current suppliers who perform well and are expected to continue to supply superior value in the long term constitute the preferred category of potential suppliers for new business.

Purchasers' Guides

Purchasers' guides, trade directories, and industrial registers are volumes that list leading suppliers and their addresses, products, branches, financial standing, and other information relevant to the buying decision. Such registers are indexed by commodity, supplier, and trade name or trademark description of the item. Examples of such guides include the *Thomas Register of American Manufacturers, McRae's Blue Book,* and *Kompass Publications* in Europe.

International Sources

Embassies and consulates of the United States in countries and cities of interest and other countries' embassies and consulates located in the United States are excellent sources of information about international suppliers. Foreign trade missions, commercial attachés, world trade centers, import brokers, and the International Federation of Purchasing and Materials Management (IFPMM) should also be considered.

Directories

In many cities, organizations publish directories listing area firms, addresses, and products sold. The yellow pages of the telephone directory are another source of information. In some cities a separate industrial section is also published.

Chamber of Commerce

The Chamber of Commerce, an association of business and professional men and women and trade and professional organizations, promotes the interest of the business community. There are over 4,000 organizations belonging to the US Chamber of Commerce, which in turn belongs to the International Chamber of Commerce.

Associations

Many local purchasing associations publish magazines, run supplier fairs, and serve as contacts for knowledge about possible sources.

Shows and Exhibits

Manufacturers, distributors, and trade organizations sponsor trade shows and exhibits on a regional or national basis to show new products and modifications of old products.

Trade Publications

Industry magazines are valuable resources with articles of interest and new product information in advertisements and editorial content.

Colleagues

Colleagues in purchasing are also good sources of supplier information. Contacts can be made through a personal network and in professional associations like the National Association of Purchasing Management and its local affiliates.

Suppliers and Salespeople

Salespeople are well informed about the capabilities and features of their own products and are also familiar with similar and competitive products. A salesperson may be able to suggest a new application for a product, eliminating the need to search for a new supplier.

Government Sources

Government agencies such as the International Trade Administration, the US Department of Commerce, the US Customs Services, and the Federal Supply Service can also be excellent sources of information.

Small, Small Disadvantaged, and Labor Surplus Area Suppliers

The Small Business Administration (SBA) has developed the Procurement Automated Source System (PASS), a computerized directory of small and small disadvantaged business sources. The directory consists of over 206,000 small, minority-owned, and women-owned businesses.

The National Minority Suppliers' Development Council (NMSDC) represents the interests of minority-owned enterprises and assists in the development of minority purchasing programs. The NMSDC has 45 minority purchasing councils in the United States. These councils are funded by over 3,000 majority-owned concerns, including more than 150 of the Fortune 500 companies.

The councils list more than 15,000 certified minority-owned concerns. The primary tasks of the purchasing councils are to certify firms as minority business suppliers (MBS); refer corporate purchasers to minority suppliers; support, develop, expand, and promote corporate minority purchasing programs; disseminate information and provide training and technical assistance; and maintain corporate and supplier directories and other publications.

Directories of socially or economically disadvantaged businesses are available from a number of sources. Published by the National Minority Business Directories, *TRY US* is the directory best known and most widely used by purchasers; it lists over 6,000 minority sources by commodity and by state. *The Guide to Obtaining Minority Business Directories,* sold by the publishers of *TRY US*, contains a comprehensive list of all minority business directories in the United States, the 45 minority purchasing councils, the 6 Minority Business Development Center regional offices, the 109 Minority Business Development Centers, and the 10 SBA regional offices.

Business Research Services, Inc., in Washington, DC, publishes two directories: the *National Directory of Minority-Owned Business Firms,* listing over 39,000 minority firms; and the *National Directory of Women-Owned Business Firms,* listing over 19,000 women-owned firms. The National Association of Women Business Owners (NAWBO) publishes a directory for its members only.

Many localities have minority chambers of commerce such as the Asian Business Association, the African-American Business Association, the US Hispanic Chamber of Commerce, or the Latin Business Association.

Many cities, counties, states, individual companies, and local minority purchasing councils conduct business fairs or trade shows for small, small disadvantaged, and women-owned businesses. Many state and local governments also compile directories of minority firms within their jurisdictions.

Supplier Information

Purchasing departments constantly receive information from potential suppliers. By cataloging, dating, and filing these mailings, the purchaser can develop an extensive base of information about possible suppliers.

Group Purchasing Organizations

Cooperative purchasing organizations also provide supplier information and services to their members. Examples of group purchasing organizations are the Educational and Institutional Cooperative Service, Inc., of the National Association of Educational Purchasers; Purchase Connections, owned by the Health Resources Institute of Los Angeles; and M.D. Buyline of Dallas, Texas.

User Departments

User departments may also have up-to-date information on suppliers or product areas. Engineering and operations or maintenance personnel in particular may be helpful. In organizations using teams, the supplier knowledge and information needed may be available from knowledgeable team members.

EVALUATION OF SUPPLIERS

Suppliers are evaluated on paper by documentation such as bid submittal, proposals, and past history. Site visits also generate additional information to help the purchaser select the supplier best able to meet current and forecast operational and strategic needs. The purchaser should evaluate the supplier's ability to perform, financial position, quality control and quality assurance, systems, organization, and management.

Plant Visits or Site Inspections

A plant visit, site inspection, or audit helps the purchaser or buying team to determine whether business should be placed with a new or previously used supplier. Approved suppliers should be visited at intervals based on their performance. If a site visit is conducted, it is beneficial to select an audit team of individuals from purchasing, quality assurance, operations, and engineering. An audit checklist will ensure that all issues are addressed.

There are a number of reasons for conducting site visits. The purchaser gets a firsthand look at the supplier's facilities, the level of technology, the education and training of the staff, the employees' attitudes toward their work, and the overall effectiveness of the supplier. If the supplier does not make a sincere effort to show its best side during this visit and cooperate fully, one can safely assume that the company probably will never do so for the purchaser.

The purchaser should weigh the costs and the benefits of a site visit. The costs may be high in terms of time, airfare, food, and lodging, but such expenditures are investments against future problems. An unaudited supplier with poor facilities, old technology, insufficient capacity, or a poorly qualified staff may result in down time or lost production as well as limited assets for future growth.

The discussions at a site visit should include operational and strategic needs of the purchaser, value improvement, and issues such as lead time, back orders, late shipments, early shipments, overshipments, quality, rejection rates, the placement of future orders, and the renewal of contractual commitments. Tours of the actual plant facilities will be particularly useful if new capacity or technology has been added.

Analyzing the Supplier's Ability to Perform

In addition to the supplier's strategic qualifications, the following operational capabilities require assessment: quality, frequency and volume of orders, length of time to process orders, delivery, quality, product and service expertise, order backlog, and productivity.

Analyzing the Supplier's Quality Assurance, Quality Control, and Related Systems

Since purchasing must ensure the quality of incoming material, analyzing the quality control side of the supplier's operation is critical. Examination of quality control procedures will reveal the supplier's attention to detail, recordkeeping, and compliance with regulations. Incoming quality control inspection procedures, documentation, and gauge calibration logs and procedures should be reviewed. All inspection procedures should be in writing, and a solid training program with periodic updates should be in place. A batch sampling technique or some form of statistical sampling procedure will probably be used instead of 100 percent inspection of incoming components.

Comprehensive shop floor quality checks should be performed regularly by the supplier's quality control staff. The audit team should review acceptance/rejection history, traceability, and calibration dates on gauges and test devices. Validation of the manufacturing process should be documented. If a statistical process control (SPC) program is in place, it should also be reviewed.

Final inspection of finished product should be done by quality control personnel. The procedure should be documented whether it is 100 percent inspection or evaluation by a statistical sample. Test methods, procedures, and instruments should be the same as those to be used by the purchaser's incoming inspection to ensure compatibility.

Frequency and Volume of Orders
The purchaser should try to find out how important the buyer's business is to the supplier. If it is a large percentage of the supplier's total business, the purchaser is more likely to get favorable treatment.

Length of Time to Process Orders
The purchaser should determine how long, on average, it takes the supplier to process orders, whether deadlines will usually be met, and how much, if any, follow-up or expediting will have to be done.

Delivery
The purchaser needs to know if the supplier has sufficient facilities and capacity to deliver the ordered materials.

Product and Service Expertise
The purchaser should know how extensive the supplier's experience is with the product, process, or service, as well as the service level provided.

Order Backlog
The purchaser should determine the company's production capacity. Is it underutilized, overutilized, operating efficiently, or backlogged?

Productivity

Productivity can be defined as the relationship between the organization's output and input. In contrast to financial ratios, productivity is a physical ratio expressed in units. Competitiveness is maintained and enhanced when productivity continually improves. Productivity improvements can be achieved through improved efficiency (same output with less input), improved effectiveness (greater output with the same input), and improved proficiency (greater output with less input). Is the supplier's productivity improving?

Analyzing the Supplier's Financial Status

The supplier's financial position should be thoroughly reviewed to determine current and future abilities to meet the purchaser's needs. A purchaser should review balance sheets, income statements, cost control history, credit ratings, annual reports, 10-K reports, financial advisory reports, Dun and Bradstreet reports, and ratio analysis.

Balance Sheets

Careful analysis of the balance sheet, which shows the financial position at a given point in time, tells the purchaser the current state of the company's financial condition. The balance sheet shows what the company owns (assets) and what it owes (liabilities). Assets are the money a company has, money it is owed, and property it owns. Liabilities are debts, including bonds and notes.

Income Statement

The income statement shows earnings for a period of time, usually one year, and reconciles sales receipts against the cost of goods sold to produce net income. The basic formula for an income statement is as follows:

Gross sales − Returned goods = Net sales

Net sales − Cost of goods sold = Gross profit

Gross profit − Total expenses = Net income before taxes

Cost Control History

The purchaser should review cost control and cost reduction steps taken by the supplier and look for areas where costs could be reduced or eliminated.

Credit Ratings

Credit ratings provide a picture of the supplier's financial history and clues to its future viability. Excessively large debt, large accounts payable, and a poor cash flow would suggest an inability to secure loans and, perhaps, an inability to generate the resources necessary to meet the purchaser's needs.

Annual Reports

All publicly held companies generate an annual report at the end of the fiscal year. The annual report includes balance sheets, income statements, and other statements of financial performance such as retained earnings and cash flow for the past three or four years. It also includes information on the company's outlook, new products and new developments, outstanding stock, acquisitions, and debt.

10-K Reports

A 10-K report provides more detailed financial information about a publicly held company than an annual report.

Financial Advisory Reports

The business section of the public library has information on the industry in which each supplier operates, the supplier's industry ranking, and the short- and long-term factors affecting the industry.

Dun and Bradstreet Reports

Dun and Bradstreet Reports provide credit information about the supplier's payment history—whether payments were made in full, whether they were on time, company history, and the educational level and experience of key management staff. These reports are often limited by the information that a company is willing to provide.

Other Factors
Using the information from the balance sheet and income statement, the purchaser can conduct ratio analysis to calculate profit ratios, leverage ratios, returns on sales, and earnings per share ratios. These ratios can be compared to those of other firms in the industry.

Analyzing the Supplier's Organization and Management

The purchaser should also analyze the supplier's organization and management, including top management commitment, stability and technical competence, training, equipment capabilities, general reputation and industry status, customer commitment, flexibility, EEO commitment, and subcontractor management.

Top Management Commitment
The purchaser should examine the organization's structure, qualifications of managers, management controls, corporate strategies, and policies and procedures. Is top management committed to customer satisfaction, quality, continuous improvement, and other initiatives to assure long-term value for its customers? Essentially the purchaser should determine if the management is a corporate strength or weakness.

Stability and Technical Competence
The company's technical competence and stability is indicated by its personnel. The educational background, previous work experience, and length of employment of the employees will demonstrate technical competence and the stability of the work force.

Training and Development
A well-documented training program indicates management commitment to being an industry leader.

Equipment Capabilities
Along with highly skilled and trained employees, the supplier should have state-of-the art equipment if it is to survive as an industry leader.

General Reputation and Industry Status
The purchaser can check in business publications on the supplier's standing in the industry. General reputation can be determined by checking references.

Customer Commitment
The degree to which the supplier is committed to its customers is determined by its willingness to commit services and support and the lengths to which the supplier goes to developing the purchaser-supplier relationship. Is the supplier supportive of long-term relationships or partnering agreements?

Flexibility
If the buyer-supplier relationship is to be dynamic and responsive to marketplace shifts, flexibility is required. Operationally, a flexible organization is not bureaucratic and is capable of frequent adjustments. Mentally, flexibility is a mindset that is customer oriented and welcomes change as an opportunity, not a threat.

EEO and Environmental Commitment
The purchaser should beware of suppliers that have violated employee discrimination or environmental laws because this may taint the reputation and image of the buying organization.

Subcontractor Management
If the supplier spends a considerable amount on raw materials and parts with subcontractors or outside suppliers, the purchaser should review the procurement system, organization, procedures, and the audit procedures that will be used with subcontractors.

Analyzing the Supplier's Labor Status

The purchaser should be concerned with three areas of a supplier's labor: employee skills, unionization, and contract expiration date.

Employee Skills
The purchaser should try to determine if the work force is stable, and if employees have the skills, training, experience, and competence to meet the purchaser's needs. The supplier's commitment to employee training and development may indicate its willingness and ability to head off potential problems.

Unionization
The existence and role of unions vary. For many organizations unionization is an accepted employee practice. Elsewhere the relationship between management and labor is such that there is no need for unions. In some organizations, the management-union relationship is one of distrust and is adversarial.

Unions are implemented in several ways. In a union shop an employee is required to join the union within 30 days following employment. A closed shop, which was outlawed by the Taft-Hartley Act of 1947, prohibits an employer from hiring nonunion employees. Some states with "right-to-work" laws have outlawed the union shop but have approved of the agency shop, which requires nonunion members to pay union dues and fees.

The contract expiration date is one of the most important things for the purchaser to know about a unionized firm because this may significantly affect the ability of the supplier to meet delivery schedules. The purchaser should gather information about the last contract negotiation, including whether there was a strike. If so, important information would include how long the strike lasted, whether there was violence or picketing that prevented deliveries, whether production continued during the strike, whether there was an inventory buildup prior to the strike, whether there was damage to the facility, the results of the negotiation, and the type of backup plan the company has in mind in case of another strike.

Analyzing Supplier Performance

Analyzing supplier performance usually entails the establishment of performance rating standards, which the purchaser can use to distinguish between good and marginal suppliers. Performance standards typically reflect the cost purchasers incur, supplier by supplier, in satisfying requirements. Past experience with a supplier's price, delivery, quality, and service is important for analyzing supplier performance.

Price

The evaluation of price in its simplest form entails comparing the price quoted against the prices quoted by other sources. Consistency of success and integrity in pricing behavior also provide measurement criteria. Cost analysis provides a more detailed analysis of the cost elements in the price and may result in a better evaluation and comparison of price quotes.

Delivery

To analyze supplier performance related to delivery, the purchaser must estimate the costs of improper delivery. A weighted cost factor can then be assigned to the different delivery possibilities, from two weeks or more ahead of schedule to one or more weeks late. Costs that enter into the standard cost base from which the weights are derived could be the expense of extra inventory, space and handling, idle operations time, the replanning caused by delinquent shipments, and any travel, telephone, and living expenses creating by having to expedite the delivery.

Quality

Product quality is often measured by the percentage of defective items purchased from a given supplier, multiplied by a disposition factor. The number of defective items purchased can be determined by sampling, and the disposition factor is based on the cost of disposing of defective material. Disposing of the defect may mean scrapping the material and reordering, returning the material and reordering, reworking the material, or even using it as is.

The cost of dealing with defective material varies with the disposition method chosen. If the defective item is used, the cost may include writing a quality data report and possibly holding a conference with the quality control engineer, the project engineer, the incoming materials inspector, and the supplier. Scrapping and reordering costs may include writing a quality report, sending a letter to the supplier, disposing of the defective materials, issuance of a second purchase order, and a second incoming inspection. Returning and reordering may involve the cost of writing a quality report and a letter to the supplier, a shipping notice, packaging of the material, issuance of a replacement order, and a second incoming inspection.

If the supplier agrees to pay for the rework, the purchaser may bear only the cost of billing the supplier. If the supplier refuses to pay for the rework, the purchaser may bear the costs of writing a quality report, scheduling the rework, and the direct labor and overhead costs to perform the rework.

Service

Service factors include the promptness with which rejected material, requests for quotations, and acknowledgements are handled. Cooperation and assistance in problem resolution should be included in the service evaluation. Areas that result in a negative rating include unplanned visits to the supplier to resolve problems, incomplete shipments, delays in inspection or receiving, and unstable financial condition of the supplier. Service evaluation is primarily based on judgment. Positive service factors include responsiveness to special requests, high quality of technical and training assistance, and voluntary suggestions for value improvement.

SUPPLIER PERFORMANCE RATING METHODS

The three principal formal techniques for rating suppliers are the categorical method, the weighted-point method, and the cost-ratio method.

Categorical Method

The categorical method is a qualitative approach to supplier evaluation. The supplier keeps a record of all suppliers and their products and services. The purchaser establishes a list of factors and then grades each supplier, usually on a scale of plus, minus, and neutral. Departments or personnel involved with the supplier also grade the supplier. The evaluators review the ratings in periodic evaluation meetings. Each supplier is then assigned an overall group rating, and suppliers are usually notified of their ratings.

Weighted-Point Method

The weighted-point method is a quantitative approach to supplier evaluation. Each performance factor is assigned a weight that reflects the purchaser's judgment about the relative importance of the specific performance factor. For example, quality might rate 50 percent, delivery 25 percent, price 15 percent, and service 10 percent for a particular purchase. Before assessing any suppliers, the purchaser must develop a formula or procedure for measuring actual supplier performance.

Each supplier's overall rating is determined by multiplying each factor weight by the corresponding performance number and then adding the results. This method minimizes subjective evaluation, but it can be used with the categorical method to include qualitative factors as well.

Cost-Ratio Method

The cost-ratio method relates all identifiable purchasing and handling costs to the value of each shipment received from individual suppliers. The lower the ratio of costs to shipments, the higher the rating for the supplier; conversely, the higher the ratio of costs to shipments, the lower the rating.

The steps normally taken in the cost-ratio method are as follows:

1. Determine the costs for quality, delivery, and service for each supplier.

2. For a given order, convert these costs to a "cost ratio" expressing the cost as a percentage of total order cost for each category.

3. Compute each supplier's overall cost ratio.

4. Adjust each supplier's price by the following formula:
    Adjusted price = Price × (1 + Overall cost ratio)

5. Use the adjusted price as a basis for evaluating supplier performance.

WHEN TO EVALUATE SUPPLIERS

Some organizations prefer to evaluate new or potential sources of supply before requesting them to bid. Others prefer to evaluate only after bids have been received and serious consideration is given to awarding the contract to one or more suppliers. Of course, existing suppliers can be evaluated while the contract is in progress. A detailed evaluation of a potential supplier is expensive, as is supplier switching. That is why many organizations are reluctant to have substantial supplier turnover.

THE OVERALL PERSPECTIVE

In the evaluation of potential sources, it is useful to keep an overview perspective. "Is this the kind of supplier we would like to do business with for a long time?" is the key question that needs to be addressed in the search stage. "Does this supplier have the potential to become an excellent supplier?" becomes the second question. The next question logically follows: "What needs to be done by the supplier and by our organization to achieve full potential?"

KEY POINTS

1.   The supplier selection decision is the most critical decision to which purchasing contributes.

2.   The supplier selection decision starts with the identification of potential suppliers.

3.   Strategic sourcing considerations include the following options: (1) Manufacturers versus resellers; (2) Small versus medium versus large suppliers; (3) Single versus multiple sources for the same requirement; (4) Local versus national versus international suppliers; (5) Traditional relationships versus preferred, certified, partners or strategic alliances; (6) Short-term versus long term perspective; (7) Creating new sources versus existing suppliers.

4. Information sources about potential suppliers include guides, directories, associations, colleagues, sales representatives, governments, and actual experience with existing suppliers.

5. Special directories provide sourcing information about women-owned and minority businesses.

6. Potential suppliers need to be evaluated on the criteria that reflect the purchaser's operational and strategic needs. Operational needs include quality, delivery, price, quantity, and service. Strategic needs include fit with the purchaser's strategic requirements. This fit may include anything that will be superior to what other purchasers may be able to obtain from the same suppliers.

7. A careful assessment of the supplier's short- and long-term suitability as a potential source will address the financial, technical, and human and management resources of the supplier. It is important that the supplier's values be congruent with the values of the purchasing organization.

8. Visits by the purchasing team to the supplier's organization and plants are essential for proper evaluation.

9. The three principal methods of supplier rating are the categorical, weighted-point, and cost-ratio methods.

CHAPTER 4

REQUESTS FOR BIDS, PROPOSALS, AND INFORMATION

Once the need of the purchaser has been determined and described and potential suppliers have been identified and evaluated, requests for bids or quotations, proposals, or information communicate the buyer's requirements to the supplier. This chapter starts with some basic solicitation concepts and concludes with a discussion of whether to use bids or negotiation.

BID SOLICITATION CONCEPTS

In competitive bidding the standard expectation is that price will be one of the major factors in awarding the contract. Two key concepts that need to be recognized in order to have a successful bidding experience are (1) comparability and (2) fairness.

To provide a common baseline for evaluating bids, all suppliers should receive the same information in the original quotation package. If additional information is provided to suppliers during the quotation process, all quoting suppliers should also receive that information at the same time.

The competitive bidding process assumes that the business will be awarded to the supplier providing the lowest price. It is critical, therefore, to confirm before bids are issued that all suppliers receiving a bid are equally qualified to receive the business. In situations where business is not awarded to the lowest bidder, a memo explaining the decision criteria should accompany the quote in the file. All suppliers quoting should be notified whether they received the job. Confidential information submitted by a supplier as part of a quotation response should remain confidential between the buyer and the supplier.

In the private sector bid price and bid content information are often kept confidential. In the public sector it is assumed that public bid documents become part of the public domain.

Bid Format and Context

Standard information for a typical bid should (at a minimum) include:

1. A description of the item to be quoted (for example, the specification, part number and description, a description of the equipment, or a definition of the services).
2. The quality required and the testing procedure to be used.
3. The quantity to be quoted.
4. The period during which the contract will be in force.
5. A list of the applicable drawings and specifications. If these are not readily available to the bidders, a copy is included with each initial bid package.
6. Where and when the items are to be shipped or installed, or where and when the services are to be performed.
7. The due date of the quotation.
8. The name of the buyer issuing the quotation, along with a phone number. If the item is highly technical, it may be prudent to include the name and telephone number of a technical contact (such as a product engineer or manufacturing engineer) who is knowledgeable about the item.

Additional information should include terms and conditions such as contract acceptance, delivery performance, contract termination, shipment rejection, assignment and subcontracting, patent rights, and payment procedures.

Request for Bid, Quotation, or Information

Information from suppliers can be collected in a variety of ways. Informally, telephone calls are the easiest and most frequently used format. Formally, a request for information (RFI) can range from information about technical specifications to an invitation to bid. Even though an RFI should theoretically be different in purpose from a request for a quotation (RFQ) or a request for a proposal (RFP), in many organizations these terms are used interchangeably.

Invitations to Bid

A request for a quotation (RFQ), an invitation to bid or a request for a bid (RFB), or a request for a tender (RFT) all attempt to solicit a quotation from a supplier. Normally, the specifications tend to be well-defined, and the price quoted will be one of the significant factors affecting supplier choice, although by no means the only one.

Responding to a buyer's requests of any kind represents a cost to the supplier. Buyers have to be aware of this. Requesting information or bids from suppliers who, in the opinion of the buyer, have no chance of obtaining business is a questionable practice. Some purchasing organizations have rules that suppliers who fail to quote three or more times in a row will be removed from the bidder's list for one or two years. This puts extra pressure on suppliers to quote even when they don't expect to be awarded the contract.

Types of Solicitations

Bid, quotation, or tender requests can be informal or formal. Informal bids are often used to solicit budget information from suppliers for estimating purposes. Suppliers have to understand that their quotation is used for budgetary or estimating purposes and may change at a later date. In construction, for example, subcontractors need to provide cost estimates so that the prime contractor can submit an overall bid. Often, once the construction contract has been awarded, the prime contractor may shop around for better deals or try to negotiate the subcontractor into lower quotes, even when the specifications have not changed.

Informal bids, often by telephone, are also used for small-value or emergency orders where the cost or time associated with an alternative solicitation approach is not justified.

Competitive Bidding

The most commonly used form of bid is the competitive bid where the request for quotation is sent to a number of different suppliers. The principle behind this is that the supplier most anxious or most capable of meeting the buyer's requirements will bid most competitively. Also, the need for an audit trail (particularly relevant in public and larger organizations) allows the bid record to speak for itself. Care needs to be taken to ensure that an appropriate cross section of qualified suppliers gets a chance to quote.

Given the repetitive nature of the quotation process and the amounts of funds expended, most organizations have formal policies and procedures to guide the bid process. In the public sector, most audits of the acquisition process tend to be compliance or conformance audits to ensure that prescribed policies and procedures were adhered to. Both the taxpayer and the supplier should be assured of fair and equitable treatment in the expenditure of public funds; and the same concerns exist in the private sector. Therefore, the process of bid solicitation, analysis, and award needs to be designed with great care to ensure that both strategic and operational needs of the organization will be satisfied in the short and the long term.

Sealed Bids and Formal Advertising

Many public organizations have a policy requiring that all contracts over a certain dollar amount be advertised. Some public agencies have instituted electronic boards for announcing bidding opportunities. These have been effectively used by the federal government in Canada, for example, to encourage smaller regional suppliers across the country to participate in the federal procurement process.

Sealed bids have to be submitted by a specified date and time. They are put into a bidder's box that will be publicly opened at a specified date and time, often right at the close of the bid deadline. The key details of the bid, name of the bidder, and price quoted will be entered into a bid log; anyone can be present at the bid opening and find out where each supplier stands. Normally, this opening is followed by a period of bid analysis to ensure that each bid conforms to the requirements, both technically as well as legally. For example, have the appropriate officers of the supplier's company signed the bid? Is the bid and performance bond documentation acceptable? Then, at a suitable future date, the supplier deemed to have the best bid will be awarded the contract.

Notifying the unsuccessful bidders is optional. It is generally considered sound procurement practice to notify all bidders, although the reasons for nonaward may or may not be disclosed depending upon the policies of the buying organization. The normal assumption with sealed bids is that no negotiation will take place before the contract award is made.

Restricted Bidding and Competition

It is clearly the prerogative of the buying organization to restrict the list of potential bidders. When the cost of preparing a bid is high, when few suppliers are capable of meeting the buyer's requirements, and when regulatory or strategic considerations dominate, such restrictions will be self-evident. Occasionally, the purchaser may wish to avoid the work of having to deal with numerous bids, or the amount of business may not justify putting a large number of suppliers to the expense of preparing bids.

Local preference laws may limit potential competition by requiring domestic or local content. Careful analysis of restricted bids by a multifunctional team, using benchmark data from comparable requirements, should be undertaken to ensure that the bids submitted are competitive.

Two-Step Bidding

Two-step bidding is normally used when inadequate specifications preclude the initial use of competitive bidding. In the first step, bids are requested only for technical proposals, without any price. The purchaser may actually pay for these bids if their preparation is expensive. In the second step, bids are sent only to those suppliers who submitted acceptable technical proposals; they are now asked to submit prices.

Requests for Proposals (RFPs)

The supplier's ingenuity can often be relied upon to provide better alternatives to meet the buyer's requirements. Many bids invite the supplier to make suggestions representing better value. Most purchasers still expect the supplier to quote on the original specifications and to indicate how the quote would be changed if the supplier's suggestion is followed.

The request for proposal (RFP) is a formal attempt by the purchaser to encourage the supplier to meet the purchaser's functional requirements without limiting the options too severely. An architect's competition for a high-profile public building such as a museum, opera house, or city hall could be likened to an RFP. RFPs are particularly appropriate where the purchaser has flexibility in terms of how a requirement may be met, sufficient lead time, and the capability of assessing different approaches, and where the purchaser suspects that alternative approaches may yield substantial value rewards.

Soliciting Proposals[1]

The traditional manual system of copying, folding, and mailing requests for proposals was seen in the past as an unavoidable and necessary part of the purchasing operating budget. For a public purchasing organization where all interested suppliers must be included on the bid list, the costs can be exorbitant. Recent advances in technology, including the proliferation of computers and facsimile machines, allow purchasers to change the system for issuing RFPs and drastically reduce the costs.

The state of Oregon's purchasing department has developed a system called the Vendor Information Program (VIP), which allows suppliers with an IBM-compatible computer and a modem to access current and historical proposal information 24 hours a day, seven days a week. Within the first year Oregon saved more than $60,000 by eliminating the paper and postage used to send out RFPs in the manual system, increased the number of registered suppliers to 8,000, and, in 1992, saved more than $12.8 million over prices paid in 1991. Surveys indicate an extremely high level of satisfaction among suppliers.

EARLY SUPPLIER INVOLVEMENT (ESI)

Early supplier involvement (ESI) is vital in situations where new product introductions are quality, time, or cost sensitive. Working closely with a supplier from the product or service concept to actual production allows

---

[1]"RFP -- ASAP!" *Napm Insights*, July 1993, pp. 40-41.

the designers to tap additional resources to ensure design for manufacturability (DFM) as well as design for procurability (DFP) and design to target quality, cost, and time. This kind of close purchaser-supplier cooperation on products of high strategic impact often precludes the standard bidding process. Standard bidding is often initiated only after the design and the manufacturing process and plans have been fixed. In contrast, under ESI both purchaser and supplier work to meet specific quality, time, and cost requirements and negotiate their way to a specific price based on mutual understanding of the relative costs.

REVERSE MARKETING

There is another way of obtaining supplier agreement to price and other requirements. In reverse marketing the purchaser makes the proposal, normally to one or a few suppliers carefully targeted for their potential to meet the purchaser's needs. In reverse marketing the purchaser decides that substantially superior results can be obtained by avoiding the standard bidding process. The need for reverse marketing is obvious when no known supplier exists for a particular requirement. It is equally applicable in strategic situations where the purchaser is seeking a competitive advantage, or in operational situations where purchaser initiative will be required to overcome supplier difficulties.

WRITING RFIs, RFQs, AND RFPs

There is no single best way to write an RFQ, RFI, or RFP. The structure, contents, and method of managing the information before, during, and after the supplier's submission of the bid should be guided by the principles of ethical and fair purchasing practice. It is also important to understand the business culture in the country and industry—some forms of bidding may not be appropriate in all places. Otherwise, the imagination of the buyer is the only restriction on the innovativeness of the proposal.

CONFERENCES WITH POTENTIAL SUPPLIERS

Conferences with suppliers before solicitation, bids, or proposals may have considerable merit under certain circumstances. These supplier conferences may be used to stir supplier interest, obtain assistance in solidifying the specifications, or help the suppliers understand the complexities and special requirements associated with the bid. Generally, such conferences are used for unique, high-value, and complex requirements.

Suppliers may be invited singly, sequentially, or as a group. Typically, the purchaser fields a team composed of purchasing as well as technical representatives (such as product, manufacturing, or quality engineering) and possibly users, marketing, or financial experts.

Topics addressed at pre-bid meetings include blueprints and specifications, quotation due dates, terms and conditions of quotation, delivery schedules and materials, releasing procedures, invoicing procedures and documentation (including incentives), requirements for awarded business (such as reporting, insurance, background checks, security clearances, and permits), and other buyer and supplier requirements.

Potential Problems

The principal danger in holding presolicitation conferences is that specifications may be slanted toward the few suppliers that attend the conference, thereby eliminating potential viable alternatives. A potential disadvantage to the pre-bid meeting is the time it takes the buyer's organization to meet with each supplier that will be quoting. Some companies have experimented with inviting multiple, even competing, suppliers to the same pre-bid conference to ensure that all parties receive the same information at the same time. This tactic has been shown to be successful, particularly in the purchase of capital equipment.

Establishing Bid Dates

One of the key pieces of information in the RFQ, RFI, or RFP is the due date of the quotation. One needs to consider the complexity of the bid, the timing constraints within the buying organization, and the number of bids concurrently issued to the same group of suppliers. Bid due dates should be realistic and enforceable.

Cancellation of Solicitations

If the buying organization decides to cancel an in-process RFQ, RFI, or RFP, each quoting supplier should be notified as quickly as possible to avoid unnecessary work for the supplier.

BID OR NEGOTIATE?

The decision to bid, negotiate, or use other means of acquisition is obviously dependent on a number of different factors.

Strategic Impact

The very first consideration is whether the particular acquisition has an impact on the organization. Is it strategic? Is it operational? Will it affect the ability of the organization to serve its external customers better? Could it affect the bottom line or the financial health of the organization significantly? Could it affect the competitiveness or the short- and long-term survival of the organization? Could it have social, political, or environmental impact? Does it fit with major initiatives, programs, or strategies in progress or planned for the organization? It has been frequently argued that purchasing's insensitivity toward exactly these types of questions has resulted in its lack of status and appreciation within its own organization. Obviously, the greater the potential impact of the acquisition under consideration, the greater the need for care and attention on the part of purchasing. Many of the bidding processes developed in purchasing have been geared to assuring that minimal purchasing time and cost are incurred, or that the most competitive price is obtained, or that an audit trail is in place. Minimizing acquisition cost or time should not be the prime consideration in strategic requirements—maximizing the strategic value should be the aim. Thus, standard bidding routines are unlikely to suffice for strategic requirements.

Strategic alliances, preferred supplier relationships, reverse marketing efforts, long-term contracts, and single sourcing should all be potential options for strategic requirements.

Other Factors

The nature of the requirement beyond its strategic impact may influence the acquisition process choice. When specifications are standard and clear, competitive bidding is relatively easy. Unique or uncertain specifications require negotiation. If special tooling or setup costs are involved, allocation of such costs and title to the tooling may have to be negotiated.

In instances where a supplier is currently supplying a product that is being revised, it is common to ask only the current supplier to quote. If this is done, there should be fundamental reasons for assuming that this supplier will be more competitive than others by virtue of its prior experience.

The timing of the requirement may determine whether sufficient time can be allowed for a competitive bidding process. This is not meant as an encouragement for requisitioners to requisition at the last minute. Obviously, proper organizational planning should allow for sufficient time to use the preferred method of procurement. Shortage of time almost invariably becomes an obstacle to obtaining superior value. The standard bidding process normally takes at least 30 days and may require substantially more for special contracts.

Long—term contracts, highly complex requirements, and high-volume contracts are normally better negotiated. On the other hand, very low-value or low-volume requirements may not be worth the expense of an extensive bidding process; telephone quotes, direct user purchases, or credit card purchases may be better. If considerations like quality, delivery, and service become important contract factors in addition to price, negotiation may again have the edge.

External Factors

External factors such as market conditions, buyers' or sellers' markets, the number of suppliers available, and the interest of suppliers may also affect the use of bidding, negotiation, or other means of acquisition.

BIDS FIRST, NEGOTIATE NEXT VERSUS FIRM BIDDING

Many organizations use bids, quotations, or proposals as a first step in supplier selection. The bids or proposals received are then analyzed, and a decision is taken to negotiate further with a subset of the suppliers who quoted. This is a sharp contrast to the practice of firm bidding, where no renegotiation is practiced. The suppliers who are asked to quote should understand whether their bids are final or may be negotiated further. Final bids are more common in the public sector. Proponents of final bidding argue that this practice is fairer and quicker and forces suppliers to make their best offer immediately. Negotiation after the first round is best defended for complex or long-term agreements or for requests for proposals where a number of items still need to be resolved after the first round of bids.

KEY POINTS

1. Requests for bids (RFBs), requests for quotations (RFQs), requests for proposals (RFPs), and requests for information (RFIs) are ways in which purchasers communicate their needs to potential suppliers.

2. Information from suppliers may be sought informally or formally. Informal requests can be made by telephone and normally cover small-value or emergency orders.

3. Competitive bidding requires that all suppliers' bids will be comparable and treated fairly.

4. In the public sector, bidding regulations often require sealed bids and formal advertising for contracts over a stipulated minimum dollar value.

5. Early supplier involvement attempts to reduce the time from identification of purchaser need to supplier delivery by consultation with suppliers at the need determination stage.

6.   Reverse marketing requires a purchaser to persuade a supplier to meet the purchaser's strategic and operational needs. Therefore, it is a proposal from the purchaser to the supplier and not a request for a bid.

7.   Conferences with suppliers may be helpful to clarify the purchaser's needs to potential suppliers.

8.   The decision whether to bid or negotiate (or both) is dependent on strategic factors and nonstrategic factors.

9.   Many organizations, especially public institutions, practice firm bidding where the supplier's bid is final. Others use the first round of bidding to establish a short list of potential suppliers for follow-up negotiations.

CHAPTER 5

BID, PRICE AND COST ANALYSIS, AND MAKE OR BUY

The evaluation of supplier quotations to determine the best overall competitive offering for a product or service is one of the primary responsibilities of the purchasing function. Each situation can be handled in a unique manner best suited to the circumstances at hand, but it is sound practice to adopt an overall framework that fits the purchaser's operational and strategic needs. This chapter discusses price and cost analysis, and the decisions to make or buy or to lease or buy.

RECEIVING, CONTROLLING, AND ABSTRACTING OFFERS

As quotes are received, the buyer should log them according to the date received and file them with the original bid package. When all of the quotes have been received against a specific bid, or at the bid due date, the buyer should prepare a brief recap of the quotations, noting the salient information in a comparison chart, and identify the supplier most likely to be selected.

In the public sector, with sealed bids, the bid log is prepared at the time of public opening. No analysis can be done at that time.

Problems Related to the Solicitation and Receipt of Offers

During the course of competitive quoting, problems may arise that require action. In the public sector, the regulations covering bidding situations may be stricter than in the private sector. Even in the private sector, however, it is useful to have and to adhere to clear policies on these issues:

1. *Time extensions and amendments to solicitation.* If one supplier is granted a time extension to respond to a bid, all suppliers must be notified at the same time that they are granted the same extension. Changes to the original bid must be communicated in a consistent and timely manner to all suppliers bidding. Significant bid request changes probably will require a bid deadline extension.

2. *Late bids (without time extensions).* The buyer must make clear the policy on late bids. The practice of not accepting late bids or returning them unopened is common today.

3. *Offers with errors, irregularities, or omissions.* If a supplier identifies a mistake in a bid after submission, common practice in the private sector allows that supplier to cancel or withdraw the bid. Courts allow the withdrawal of bids only after determining (1) that the mistake was mechanical or clerical in nature, not an error in judgment, and (2) that the bidder was not guilty of negligence in making the error or in delaying to notify the buyer of the error. In the public sector, bid bonds protect the purchaser from low bidder withdrawal. Caution should be taken to ensure that a supplier does not repeat these types of mistakes.

    If an error, irregularity, or omission is so out of proportion as to indicate a mistake, the buyer should seek confirmation of the bid from the bidder before proceeding with the award. If a mistake is confirmed, the bidder should be allowed to cancel or withdraw the bid without penalty.

4. *Conflicts of interest.* Fundamental ethical practice requires that no employee who has any authority to purchase goods or services or is in a position to influence purchase decisions in any way should be employed by, hold any position with, serve as a director of, have a financial interest in, or have a business relationship with any outside concern that is a supplier of goods and services to the buying company.

BONDS

Bonds afford buyers an opportunity to insulate their organizations from the consequences of seller failure to perform or to pay for labor and materials. They are common in the construction industry. The public sector often requires bid and performance bonds at the time of bid submission and will require the supplier who is awarded the contract to post the necessary bonds. Common types of bonds include:

- *Bid bonds.* A bid bond guarantees that the bidder will accept the contract for the stated bid amount, or that by virtue of the bidder's failure to so perform, a buyer who has to pay more to have the contract performed will be compensated for the difference.

- *Labor and materials bonds.* If a contractor fails to pay for labor or materials, damaged subcontractors or suppliers can file a mechanic's lien against the property, thereby preventing its use by the buyer/owner. Under this kind of bond, the bonding company guarantees payment for labor and materials, which insulates the buyer/owner from such risk.

- *Performance bonds.* Under this kind of bond, the bonding company guarantees damages to a buyer/owner in the event of a seller/contractor's failure to perform as agreed.

A supplier whose bonding company is called upon to make good for supplier default on a contract will have difficulty obtaining bonds in the future. This may prevent the supplier from bidding on future contracts. The alternative of setting aside a sum of cash in the amount of the bond that would otherwise have been used is very expensive. The threat of calling a bond is, therefore, a powerful persuader for a supplier to perform. Bonds are seldom called. Since the bonding company charges a fee for its service, the contract price is understood to include the cost of bonding.

Even when a bond has to be called, the purchaser faces a situation in which the supplier's nonperformance has delayed delivery or caused other difficulties that need to be resolved. The financial compensation from the bond usually does not solve the supply problems that have been created. Therefore, the existence of a bond should not be seen as an opportunity to reduce purchaser vigilance in supplier selection or contract administration.

Nonresponsive or Late Bids and Offers

Buyers have the right to require that all bids be submitted by the due date in order to be considered. However, two practices need to be observed when this is the stated policy. First, the buyer must not accept late quotations from any supplier. Second, a reasonable amount of time must be allowed for suppliers to respond to the quotation in a timely manner. This is based on the judgment of the purchasing organization, with past experience for the specific commodity or service being used as a guideline. If for extraordinary reasons the bid due date is extended, all suppliers should be notified of the new bid closing date.

OFFER RESPONSIVENESS

Specifications and Statement of Work

Perhaps the most critical aspect in evaluating bids offered by suppliers is to ensure that the quotation, or the statement of work within the bid response, meets the specifications and requirements of the initial RFQ, RFP, or RFI package. This is important for two reasons.

First, one needs to ensure that the supplier will meet the specifications, requirements, or statement of work initially defined. If this is not the case, the buyer must be aware of the implications in accepting the bid. Not meeting specifications or submitting an alternate statement of work can have either a positive or negative impact on the buyer's organization. At worst, it may cause quality concerns or it may not meet the minimum requirements defined by the buying organization. The buyer must ensure that (1) suppliers are quoting to specification and (2) the quoted exceptions are acceptable to the total organization.

Second, to evaluate all quotes fairly, suppliers must quote based on a comparable set of specifications or a comparable statement of work. Otherwise, it is extremely difficult to determine which supplier offers the most competitive package.

Quality Requirements and Other Terms and Conditions

As with the base specifications or statement of work, suppliers must be aware of the quality and testing requirements and other terms and conditions for the goods or services that are part of the initial RFQ, RFP, or RFI package. Suppliers must acknowledge that they will meet these requirements. If suppliers quote exceptions, it is the buyer's responsibility to ensure that the organization can accept these alternate proposals.

Technical Proposals and Presale Technical Service

Presale technical service is offered by some companies as a part of the quotation process, particularly when technical products or services are being purchased. Purchasing must ensure that it does not take unfair advantage of suppliers offering presale technical assistance, but at the same time must ensure that it receives all of the assistance to which it is entitled prior to award. Acceptance of more presale service than is customary in the industry may obligate the purchaser to more than is anticipated.

Technical and Operational Analysis

Purchasing should actively involve the requisitioning, engineering, manufacturing, materials control, and other using departments in the bid evaluation process. This is a natural extension of involving these departments in the initial definition of specifications or the statement of work. Each department should have a guideline by which it evaluates supplier proposals and should be made aware of fair practice procedures before beginning the evaluation process.

COST/PRICE ANALYSIS

Often done exclusively by purchasing, but sometimes involving finance or engineering, cost/price analysis is frequently the core of the bid evaluation process.

Price Determination

Prices may be determined in a number of ways. The buyer may use price analysis or cost analysis to decide what he or she considers a fair and reasonable price. The buyer can perform cost and price analysis using a number of techniques.

The seller has two basic methods of setting a price: the cost approach and the market approach. In the cost approach the seller sets a price to cover direct costs, a portion of overhead and indirect costs, and profit. The buyer may seek a low-cost producer, work with the supplier to reduce manufacturing costs, or question the size of the margin over direct costs. Cost analysis techniques coupled with negotiation give the buyer an opportunity to influence the price.

In the market approach, prices are set in the marketplace by supply and demand and may have little relation to cost. In market-based pricing, the purchasing plan devised by the buyer may focus more on nonprice incentives such as transportation concessions, strict delivery schedules, or superior quality or service. In this environment it may be worth looking at the possibility of finding a substitute or of making rather than buying. The structure of the terms, such as longer-term contracts or closer relationships, may induce suppliers to ignore market conditions if the benefits to the supplier are great enough.

PRICE ANALYSIS

In performing price analysis the buyer may compare a seller's price to the price proposed by other sellers, or to price benchmarks such as catalog or market prices, or to historical prices. The buyer does not look at the cost elements of the price. Often price analysis may not be sufficiently thorough to ensure that the buyer obtains the best price. Some analysis of the cost elements will be required.

James Mayers, vice president of Citibank's Systems and Technology Division, tells of a national hardware supplier that took advantage of a long-term relationship by maintaining unnecessarily high price structures. A survey of competitors revealed that other suppliers could offer quicker turnaround times and longer warranty periods at better prices. Not wishing to lose the business, the original supplier met the competitor's prices and reduced its cycle time. Performance improved, Citibank saved money, and the supplier was able to quote on other business.[1]

COST ANALYSIS

Cost analysis enables the buyer to estimate the seller's costs as a means of evaluating the fairness and reasonableness of the proposed price. In a request for proposal the buyer should ask for a cost breakdown. If it is not provided or as a means of analyzing the information provided, the buyer may have to do a cost buildup.

In comparing suppliers, the buyer must understand the nature of the costs and be able to explain any discrepancies among suppliers concerning labor, material, overhead costs, and profit.

Basic Cost Concepts

Price and cost analysis requires that the person using these techniques has a thorough understanding of how costs behave in real life and how costs are accounted for. Although a thorough treatment of costs is beyond the scope of this text, it is useful to highlight a few basic cost concepts here.

Relevant and Irrelevant Costs
A distinction should be made between relevant and irrelevant costs. Relevant costs are costs that change as a result of a decision and have an impact on the decision. For example, if a buy decision allowed the organization to sell a warehouse, then the reduced operating costs and cash receipts from the sale would be relevant to the decision. If a buy

---

[1]Eberhard E. Sheuing, PhD, CPM, "The Dynamics of Lasting Links," *NAPM Insights*, May 1992, p. 13.

decision only freed up unneeded warehouse space, then the organization would not enjoy any reduction in storage costs, so these would be irrelevant to the decision.

Fixed versus Semivariable and Variable Costs
Fixed and variable costs should also be considered if they are relevant to the decision. The taxes on the warehouse might be a fixed cost, whereas the staffing costs might be semivariable and shipping costs might be totally variable.

Overhead Costs
Overhead costs are necessary for the existence and operation of activities. In an organization housed in an office building, space overhead costs might include heat, maintenance, light, power, and rent. Supervisory overhead might include managerial salaries. Support overhead could include the cost of the receptionist, computer, telephone, cafeteria, security, personnel, and accounting. Overhead costs are difficult to allocate to any one product or activity in a multiproduct organization. In traditional cost accounting, overhead costs are allocated across product lines on the basis of some predetermined percentage.

Activity-Based Costing
One of the problems with any cost reduction strategy is finding an accurate way of relating costs back to a specific product. Traditional cost accounting has used allocation bases that do not consider the degree to which indirect costs are attributable to a specific product. Activity-based costing attempts to define as many costs as possible as direct costs. For example, if a product requires extensive engineering support, then the appropriate number of hours of engineering support would be assigned to that product as costs. The only costs exempt from this application are research and development and excess capacity. Although it may be difficult to track indirect costs in such a manner, even a rough cut may lead to a better cost picture than that of traditional cost accounting.

Opportunity Costs
When an organization decides to invest its capital in equipment instead of investing in a high-return liquid investment, the organization must account

for the lost opportunity of receiving the return on the high-yield investment. The amount of money that is **not** earned because the capital is invested in machinery rather than a financial security is considered the opportunity cost.

In the private sector, the cost of capital includes the notion of opportunity cost and the risks associated with various applications of funds. In the public sector, opportunity costs are just as real. The public entity needs to deal with trade-offs such as the difficulty and cost of acquiring additional funding, the benefits of decreasing debt, and the benefits of funding certain initiatives or programs. It is naive to assume that the opportunity cost in the public sector is the current cost of borrowing or, worse yet, zero because funding comes from taxes.

Total Cost of Ownership or Life Cycle Costing

Basing a purchasing decision strictly on acquisition price may be misleading if the other costs associated with the purchase are ignored. Price is just one element of total cost. Other relevant costs may be the costs of transportation, duty, brokerage fees, poor quality, poor accounting practices, late delivery, or poor customer service. The complete financial implications of a purchase may be more accurately analyzed through a total cost model.

Life cycle costing is a total cost method used to determine the total cost of ownership over the useful life of an asset. To perform life cycle cost analysis, the buyer must first identify the operating cycle for the equipment and identify and qualify all the factors that affect costs. This analysis allows a comparison of equipment alternatives with different operating costs over the life of the equipment.

Everyone is aware that the cost of owning an automobile is far greater than the purchase price alone. Operating costs like fuel and maintenance and other costs such as insurance, licensing, and depreciation also need to be considered, as well as the disposal value of the vehicle at the time of resale or trade-in. Since not all of these costs are incurred at the same time, it is wise to use a present value model to determine the best value at the time of purchase. Life cycle costing cannot, of course, account for such intangibles as personal preference for a certain shape, style, or color. It is, nevertheless, a powerful tool to identify the relevant cost of ownership over time.

Total Cost of Ownership

At Florida Power and Light, solid-state kilowatt-hour meters were purchased to replace hybrid mechanical/electronic meters for its commercial and industrial customers. Because the new meters were more accurate, energy usage was better captured; revenue was ensured and customers were not overcharged. Engineering and purchasing were involved in the decision. The analysis of the total cost of the application of the new product or design included assessing the new technology costs, start-up costs, downtime costs, standardization costs and benefits, and product life.

New technology costs are direct costs associated with the application of the new design or product, including special engineering costs and product testing costs. For example, time investments such as engineering hours must be captured. Start-up costs refer to installation and start-up activities including the costs of field material, subcontractors, energy, special testing, special construction equipment, administrative work, overhead, supervision, start-up labor, removal labor, training, and computer software enhancements.[2]

VALUE ANALYSIS AND VALUE ENGINEERING

Two important tools in cost reduction are value analysis and value engineering. Both compare function to cost in a methodical analysis of a part or component to determine if there is a more cost-effective way to perform the function than the existing one. Book 3 in this series covers value analysis and value engineering in depth.

LEARNING CURVE AND MANUFACTURING PROGRESS FUNCTION

At the outset of a new manufacturing process or the production of a new product or service, extra time and material will probably be required in

---

[2]José Fernandez, CPM, "Cost of Product Application," *NAPM Insights*, September 1992, p. 5.

learning how to produce more efficiently. As the supplier learns the process better, costs are reduced through fewer rejects and rework, better scheduling, fewer engineering changes, possible improvements in tooling, and more efficient management systems. The *learning curve* is a way of measuring and explaining this learning process and justifying price reductions over time as the supplier becomes more efficient.

Traditionally, the learning curve has been applied to the empirical relationship between volume produced and the number of direct labor hours required for production. Each time production doubles, the amount of direct labor required, per unit, is reduced by a set percentage.

Here is an example of an 80 percent cumulative learning curve:

Unit Number	Unit Direct Labor Hours	Cumulative Direct Labor Hours	Cumulative Average Direct Labor Hours
1	1,000	1,000	1,000
2	800	1,800	900
4	640	3,142	786
8	512	5,346	668

The curve selected by an organization depends entirely upon the empirical data collected from actual experience. It is generally found that higher rates of learning, say 75 to 85 percent, apply to complex activities, while lower rates of learning, 90 to 95 percent, apply to simpler tasks. It may be necessary to estimate the rate of learning based on past experience in similar circumstances and then renegotiate based on actual data after several production runs.

Certain curves are known to apply to particular processes or industries. The learning curve has traditionally been used for purchasing complex equipment in the aircraft, electronics, and other highly technical industries.

Since the direct labor input has decreased in most organizations, the focus of efficiency improvements has expanded to include improvements in supervision, management, and materials.

Manufacturing Progress Function and Continuous Improvement

The term *manufacturing progress function* applies the learning curve concept to all of the learning and improvement efforts made by the whole organization in its efforts to achieve continuous improvement. The manufacturing progress function is also referred to as the *improvement curve* and the *organizational improvement curve* to reflect the impact of other areas. For example, through value analysis, the design of a product may be modified for ease of assembly or to reduce defects. Incoming materials may be of better quality. Equipment modification may allow for faster setup and reduced downtime.

The implication of the manufacturing progress function or the learning curve is that continuously decreasing prices and continuously improving quality are not unreasonable requirements from suppliers.

The improvement curve or manufacturing progress function may be used in scheduling production and staffing requirements over a period of time and to analyze the supplier's unit cost of production. The improvement curve can be used to develop target costs for new products; aid in establishing a starting point for pricing in a negotiation; and provide make or buy information, delivery schedules, and progress payment schedules for suppliers.

Since learning varies greatly across industries, processes, and parts, curves should be based on actual data or carefully selected past experience. Suppliers must be motivated to set improvement goals and must plan how to achieve them. Factors beyond the supplier's control such as fluctuating commodity prices may have to be dealt with separately.

As inflation decreases and competitive pressures increase, the combined effects of learning curves, manufacturing progress functions, and continuous improvement initiatives should result in continuous improvement. Thus, contracts with continually decreasing prices or with continually improving value targets over time are becoming more common.

MAKE OR BUY

The make or buy question is one of the most fundamental decisions facing an organization. Should a school system provide its own janitorial or dietary service or contract with an outside service? Should a manufacturer make a component currently sourced outside, or conversely, outsource for an item currently made in-house?

The key question that must be answered is, Is there a strategic reason or value in making over buying? Traditionally, many organizations have chosen the make option. For example, in the past, most auto manufacturers purchased raw materials and assembled their own car seats. Now, most seats are outsourced. Recent management trends, such as closer buyer-supplier relationships; emphasis on competitiveness, productivity, and flexibility; and focus on corporate strengths have led to more firms exercising the buy option. Security, food, maintenance, programming, legal, and a host of other functions traditionally performed in-house have been contracted out in many organizations.

In the public sector, the transfer of public services to private contractors and the transfer of government-owned enterprises to private systems is called privatization. For example, a municipality may decide to hire a sanitation contractor rather than maintain its own sanitation department.

The decision to form closer relationships with key suppliers in a partnering arrangement or strategic alliance can be seen as an attempt to achieve most of the benefits of vertical integration without some of the major disadvantages.

Some of the reasons an organization may decide to make rather than buy are:

1. Quantities are too small.
2. Suppliers cannot provide the processes necessary to meet quality requirements.
3. Supply is assured or there is closer coordination between supply and demand.
4. Technological secrets are protected.
5. It is cheaper to make than to buy.
6. Equipment or labor would be idle otherwise.
7. To smooth fluctuations in the operation.
8. To avoid sole source dependency.

9. There may be overriding competitive, social, environmental, or political reasons to make. An organization may be required to process a certain amount of raw materials within national boundaries. An organization may decide to make because of a sense of social conscience.

An organization may face some dangers in deciding to make rather than buy:

1. The organization lacks the production or technological expertise so the make decision would not be of strategic advantage.
2. If excess capacity exists, the organization may end up on the market competing with suppliers to the detriment of supplier relations.
3. End customers may have a preference for a branded component that makes the finished product more acceptable and marketable.
4. If the process is not a core activity, the organization may have difficulty maintaining long-term economic and technological viability. The strategic goals and objectives of the organization must be the driving force behind the decision.
5. It may be difficult to reverse a make decision once it has been implemented.
6. The long-term total cost of making may be difficult to determine. For example, the cost of labor may be underestimated, or changing economic times may drive up the cost in terms of salaries, wages, benefits, and training.
7. Inflexibility in selecting possible sources and substitute items is introduced. If special tooling or equipment is required, once the make decision is implemented, the organization may effectively cut off procurement options for the long-term.
8. Organizations may lose sight of the core activities that distinguish them from others and expend resources on activities that do not add value to the organization.
9. The organization will face the risks of changing economic conditions, such as cyclical and long-term trends in the industry, changes in demand, technological advances, and unpredictable factors such as government regulations, tax policies, and international control.

Procedure for Conducting Make or Buy Analyses

The decision to make or buy involves three steps:

1. Determine the feasibility of the idea.
2. Determine the need.
3. Determine the total cost associated with the methods or processes.

Determining Feasibility
The first step in make or buy analysis is to determine if it is in the firm's or organization's strategic interest to make the product or perform the service itself. The idea that an organization should have core competencies and avoid spreading itself too thin has been widely accepted. The second issue is whether the organization is capable of making instead of buying.

To make in a manufacturing environment requires production equipment, personnel, material, space, supervision, overhead, maintenance, taxes, insurance, and other costs. To perform a service in-house requires the personnel, training, expertise, supervision, and other costs. An inability to meet the costs and requirements or an analysis that reveals that resources would yield greater results if committed elsewhere would favor a buy decision. The buy decision involves lower investments in facilities, a smaller labor force, lower plant costs, and less overhead, taxes, insurance, and supervision.

Determining the Need
The organization must clearly describe the need as was discussed in the specification section. What is needed and in what quantities? What parts can be manufactured by the firm and what parts can be manufactured by others? What services can be provided most effectively by the firm and what services can be best provided by outside suppliers?

Determining the Total Cost of Make or Buy
After dependability of supply, total cost is a foremost concern in the make or buy analysis. Can the part or service be made or provided in-house less expensively than it can be purchased? Many of the cost issues and concepts discussed earlier relate to the make or buy decision.

A make or buy analysis must account for all costs and not just those of purchasing or operations. The organization must decide how making the item in-house, moving it to another firm, providing the service in-house, or outsourcing it will change the costs and performance of the organization. The total cost concept requires that all costs must be analyzed. These costs include direct material and labor, energy, transportation, inventory, quality, obsolescence, capital costs, and appropriate overhead charges. Buying costs include the purchase price, inbound transportation, inventory, the cost of establishing and maintaining the relationship with the supplier, and the costs of quality, payment, and disposal. Making costs include the acquisition of raw material, inbound transportation, production, direct labor, and all other costs associated with making the items in-house or performing services internally.

Care needs to be taken to include the appropriate overhead costs in the analysis. Moreover, where a change from existing practice is being contemplated, the one-time costs of switching also need to be recognized. For example, separation allowances for staff dismissed, extra inventories built up to cope with the change, training costs, and inefficiencies during start-up all need to be recognized.

Interestingly, the organization's ability to purchase effectively is an important component of the make or buy decision. An organization capable of developing its supply function and its supplier base into part of its competitive advantage should maximize its buy options.

LEASE OR BUY

The decision to lease or buy is similar to the make or buy decision in terms of the analytical process. It is normally reserved for capital assets.

A lease is a contract where one party (the lessee) has the use and possession of an asset owned by another party (the lessor) for a period of time in return for monetary payment. The lessee makes scheduled payments (usually monthly) to the lessor.

Lessees do not own the equipment but are permitted to claim rental payments made on such assets as tax deductions. Lessors make a profit on the difference between the ownership cost of the assets and the rental or lease rate. At the end of the lease term, the lessee may purchase the asset, return it to the lessor, or renew the lease, depending on the lease specifications.

The advantages of leasing are as follows:

1. Lease rentals are operating expenses and may be deductible for income tax purposes. Tax savings may be realized if lease payments exceed allowable depreciable amounts.
2. Initial outlay is small. In a buy decision, the initial capital investment is high and there are the added costs of interest charges on loans.
3. Expert service is available and the cost of maintenance is usually borne by the lessor. If the equipment is complex, maintenance may be particularly important.
4. The risk of obsolescence may be reduced or eliminated. Leases may include a provision for replacing obsolete equipment with new or updated equipment. The lessor bears responsibility for selling or disposing of unneeded or outdated equipment.
5. Short-term needs are met, such as special jobs or seasonal business. If a purchased piece of equipment is temporarily out of service, leasing or renting provides operating continuity.
6. A test period is provided before purchase or lease extension.
7. The supplier bears the burden of investment, thus freeing capital for the lessee to invest in other, more profitable ventures. Instead of a large capital outlay, the lessee makes smaller, regular payments.
8. The lease option broadens the sources of supply and increases the competitive alternatives available to the purchaser. If funds for capital expenditures are unavailable, leasing may be the answer.

Here are the disadvantages of leasing:

1. Final cost may be high. Leasing is generally a more expensive way to obtain a piece of equipment because the lessor bears all risks of ownership including inflation. Total leasing charges over the life of the equipment exceed the original purchase price.
2. The lessee is required to provide the lessor access to the equipment.
3. There is less freedom of control and use. The lessor may place restrictions on the use and operation of the equipment that the

lessee would not have if the equipment were purchased. Also, the lessee is barred from customizing the equipment.

4.   If the item appreciates over time, such as land, buildings, and airplanes do, the firm owning the asset after its effective life has an advantage. Any residual value or investment recovery on the equipment may be reaped by the lessor.
5.   Lessees must carefully review the terms of the lease because the lessor may place all the risk on the lessee.

Types of Lease Arrangements

The two major types of leases are operating and financial leases. Other types of leases include leveraged leases, master leases, dry lease/wet lease, and sale and leaseback.

Operating Leases

Operating leases are used to lease equipment when a firm has a temporary need such as a special production or maintenance job. They are usually short-term, for a fixed period less than the life of the equipment, and at a fixed financial commitment less than the purchase price of the equipment. In its basic form the operating lease is noncancellable. Service and flexibility are the key factors in the operating lease. The lessor assumes full responsibility for insurance, taxes, obsolescence, maintenance, purchase, and resale of the equipment. The lessee must evaluate the charges for these services against other available alternatives. The equipment does not have to be reported as an asset, and lease payments are shown only when they are made.

Financial Leases

Financial leases are used by the lessee to gain financial leverage and related long-term financial benefits. Financial leases run for the full life of the equipment; many are noncancellable. The lease cost includes the lessor's fee, the interest rate, and the depreciation rate of the equipment. Financial leases are shown as assets of the lessee, and the lease payments are a liability. There are two types of financial leases: full payout and partial payout.

Full payout leases require the lessee to pay the full purchase price of the equipment plus applicable maintenance, service, record-keeping, and insurance charges on a regular payment plan. *Partial payout leases* consider the residual value of the equipment at the end of the lease period; the lessee pays the difference between the purchase price and the residual value plus interest and charges.

Other Types of Leases

Leveraged leases involve a third party lessor that buys the equipment and then leases it to another firm. A *master lease,* similar to a blanket order contract, uses prenegotiated terms and conditions for certain categories of equipment. When a short-term need arises the master lease applies, and the price and length of use are then determined. Dry and wet leases originated in the aircraft industry. A *dry* or *straight lease* provides only for financing. A *wet lease* includes financing plus fuel and maintenance. In a *sale and leaseback* arrangement, the owner of a piece of equipment or property sells the asset to a second party and then leases it back. Firms often use a sale and leaseback to obtain capital.

Leasing Structures

The lessee should be concerned about the structure of the lease because the structure gives information about how the lessor will profit from the deal, and this information helps in negotiating lease terms. Four types of leasing structures are the full service lessor, the finance lease company, the captive lease company, and bank participation.

The *full service lessor* or *third party lessor* has its own source of financing, purchases the equipment from the manufacturer, and provides all services.

The *finance lease company*, which may be formed by institutional investors or wealthy individuals, provides the financing, but the lessee deals directly with the equipment manufacturer. The investors put up 10 to 20 percent of the purchase price, and financing is obtained on the remainder. The debt is secured through lease payments, and the investors are able to claim tax deductions on the entire cost of the equipment.

In *captive leasing* the original manufacturer leases the equipment. Many high technology or high-cost products are offered this way as a means to increase sales. For example, leasing large computer systems enables users to avoid high cost and potential obsolescence.

With *bank participation* the lessee uses its credit rating to attract the bank, but the lessor still deals with the purchase, service, and disposal of equipment. The bank typically puts up 15 to 30 percent of the capital, borrows the balance, and uses the lessee's payments to cover the cost of borrowing plus profit.

KEY POINTS

1.   Quotations or bids received from suppliers need to be logged, analyzed, and summarized.

2.   The most critical assessment centers on the conformance of the bid with the original RFQ, RFP, or RFI package.

3.   Requisitioners and other knowledgeable individuals or departments should be involved in the assessment of offers received.

4.   Price analysis benchmarks prices received against other price indicators.

5.   Cost analysis allows the buyer to estimate the supplier's costs to ensure that proper value will be obtained.

6.   Thorough familiarity with basic costs and accounting costs such as relevant and irrelevant costs and fixed, variable, semivariable, overhead, and opportunity costs is a prerequisite to sound cost analysis.

7.   Activity-based costing, total cost of ownership, value analysis, value engineering, and the learning curve, organizational

improvement function, or manufacturing progress function are tools to seek improvement in value.

8.   Make or buy is one of the key strategic choices made in every organization. Purchasing's involvement in this decision is essential. Purchasing's ability to buy effectively should be one of the factors influencing the make or buy decision.

9.   Leasing permits an organization to conserve its funds in acquiring assets. Therefore, the lease or buy decision has major financial implications.

10. An operating lease generally covers a temporary need while a financial lease is longer term.

CHAPTER 6

NEGOTIATION

Negotiation in the supply context has traditionally been identified as a process by which buyer and seller try to reach agreement. For purchasers, negotiation with suppliers is clearly a key activity; but negotiation inside the purchaser's organization is equally valuable. Obtaining agreement from users, requisitioners, financial people, and any others in the organization requires negotiation skills of the highest order. Changing a buying procedure, product or service, supplier, specification, quality, test, delivery date, quantity, or price may have substantial benefits, but only if others in the organization agree. Purchasing's influence with suppliers depends on the internal support behind purchasing's position. Therefore, the negotiation process identified in this chapter is equally applicable internally and externally.

Proper planning is the key to negotiation success. Certain individuals seem to be born negotiators who can achieve miracles without much preparation. Most people, however, need to plan and prepare diligently to ensure negotiation success. Preparing a written negotiation plan is an effective means to ensure appropriate preparation and planning for an upcoming negotiation.

Many of today's management initiatives outside of the supply area, such as total customer satisfaction, quality, time compression, continuous improvement, total cost of ownership reduction, employee empowerment, and reengineering, have reinforced the need for negotiation internally and with suppliers. Materials management initiatives such as MRP, JIT, supplier base reduction, early supplier involvement, preferred suppliers, single sourcing, long-term contracts, and strategic alliances also have increased the need for negotiation.

Negotiation strategies and tactics cover three levels:

Strategic planning. Strategic planning requires the negotiator to develop negotiation strategies that will optimize attainment of the overall philosophy, objectives, and strategies of the organization. Essential knowledge includes organizational values, product and market mix, customers, environmental goals, and the basic goals of the organization concerning technology, price, and policy.

Administrative planning. This refers to the logistics of getting people and information in place for the negotiations.

Tactical planning. This involves getting optimal results at the bargaining table. It involves the setting of goals, and evaluating the strengths and weaknesses of the opponents. A careful study should be made of issues, problems, agenda questions, concessions, commitments, promises, pricing, quality, and delivery performance. The team leader should be selected and negotiating tactics reviewed and discussed. Timing, strengths and weaknesses, delays, and deadlock tactics should all be planned. A buyer should gather as much pertinent data and history as possible.

FUNDAMENTAL PURPOSE OF NEGOTIATION

The fundamental purpose of negotiation is to reach agreement in such a way that after the negotiation is completed both parties will perform as agreed. The term win-win negotiation has been coined to reflect this goal. If both parties are satisfied that the agreement is acceptable economically and psychologically, it is in their own interest to commit to performance and to ensure that performance conforms to the agreement. The participative and face-to-face dimension of negotiation gives it its special power. The planning process for negotiation must be able to define common economic and psychological ground for both sides.

Reasons for Internal Negotiation

There are many reasons why purchasing should negotiate effectively internally. These may be divided into four classes:

1. Initiatives originating with external customers.

2. Initiatives originating within the organization, but outside of supply.

3. Initiatives originating within the supply area.

4. Initiatives originating with suppliers.

1. Initiatives Originating with External Customers

External customers often aggressively pursue suppliers to make changes that would benefit the external customer or its customers. Not surprisingly, many of these efforts may affect both purchasing and the supply chain. Obviously, where it is reasonably possible, attempts should be made to accommodate such requests, setting in motion negotiation sessions. There may be situations, however, where what the external customer requests is not feasible, and the negotiation turns back to the external customer to see if other alternatives may be pursued, or to persuade the external customer to accept the inevitable.

2. Initiatives Originating within the Organization, but Outside of Supply

Internal customers will frequently wish to change plans, originate new requirements, or be involved in tasks or projects that will ultimately affect both purchaser and supplier. Outsourcing services or items previously provided in-house is only one example. Purchasing's early awareness and involvement in these initiatives or changes is vital to avoid delay and to assure that the appropriate sourcing decisions can be made. Negotiations are likely to be both internal and external. The internal customer's perceptions of purchasing's willingness and ability to support their customers will affect purchasing's future effectiveness.

3. Initiatives Originating within the Supply Area

Initiatives originating within the supply area may require others in the organization as well as external customers and suppliers to adjust and accommodate. Some initiatives, such as a suggested specification change, may require additional resources from designers or engineers. Standardization, simplification, value analysis, early purchasing and supplier involvement, supplier base reduction, and EDI all require substantial input and resources to be successful. The ability to initiate supply projects and get them approved and funded is essential to the long-term viability of the supply function. Negotiating for appropriate top management, peer, and other internal involvement in supply initiatives should be a high priority.

4. Supplier Initiatives

It is natural for suppliers to approach purchasers with new and improved products and services and suggestions for value improvement. It is purchasing's role to deal with these suggestions and to enlist the help of others in the organization to evaluate, test, and implement those that prove to be beneficial. If suppliers perceive purchasers as incapable of assessing the merits of their ideas or incapable of persuading the appropriate internal decision makers, they will try to bypass the purchasing function. Ensuring that supplier initiatives get proper and prompt assessment and feedback is a good way to avoid such bypasses.

ADDITIONAL REASONS FOR NEGOTIATION

There are many additional reasons why negotiation may be the best alternative to resolve a purchasing situation, including:

- Lack of competition.
- Price, quality, delivery, and service needs.
- Buying production or service capabilities.
- High buyer or seller uncertainty.
- Urgency.
- Long supplier lead times.
- Necessity for flexible contracts.
- Lack of firm product specifications.
- Single-source strategies.

Lack of Competition

In a situation of limited competition where a small number of suppliers exert considerable control over and pressure on pricing, negotiation may be the only alternative. While it may be difficult to gain price concessions under these circumstances, it is not impossible. Through careful preparation and planning, the buyer or negotiator may identify a weakness or need in the supplier and develop a strategy for winning a price concession from the supplier. Also, if the buyer needs a large volume or a long-term arrangement, the seller may be more interested in the contract.

Price, Quality, and Service Needs

The buyer may gain price concessions by offering 100 percent of the volume (previously split between several suppliers), multiyear contracts, the promise of increased volumes (because of new product developments), and the willingness to grow and develop as a partner.

Since quality is normally not an area where compromises are possible, the buyer must ensure that the quality terms are clearly outlined in the specifications provided. A clear understanding between buyer and seller of the required quality standards and testing increases the likelihood of quality satisfaction. It will save time and money in resolving any issues about poor quality, rejects, or rework.

All the requirements for good service must be spelled out in the specifications and in the contract. Hashing out the details in the negotiation will ensure speedy resolution of problems later on. Issues to consider are 24-hour emergency numbers, deliveries on holidays and weekends, emergency shipments, engineering assistance, response to requests for early shipments, a personal representative to handle the account, and a thorough knowledge of the buyer's needs and the supplier's capabilities.

Buying Production or Service Capabilities

Specific requirements regarding production service capabilities may be needed in advance of the establishment of price. Frequency, time of day or night, quality, insurance coverage, liability, material costs, uniforms to be worn, and holiday work for the service are just a few of the items that should be reviewed for cost.

High Buyer or Seller Uncertainty

The buyer may decide to use negotiation when there is a high degree of uncertainty about the seller and the market within which the seller operates. A buyer may wish to strengthen his or her position in a period of tight supply caused by material shortage or the absence of a major producer. If the buyer is uncertain about the supplier's commitment to servicing all of the buyer's needs, the buyer may want to solidify the commitment through a binding contract and a negotiated price.

Urgency

When a buyer needs an unscheduled shipment or a rush on a critical item, negotiation may be the best means of coming to an agreement about price and other terms. The supplier may be willing and able to meet the urgent need but may also want certain concessions from the buyer in return, perhaps in the event of production delays or problems. Should a buyer go to a secondary, previously unused, supplier, the supplier may wish to negotiate additional business in return.

Long Supplier Lead Times

Long supplier lead times may require the buyer's organization to carry more inventory or safety stocks, order further in the future, and conduct more expediting activities to ensure deliveries. Through negotiations, the buyer may commit to the supplier for specific quantities for a certain time period. This enables the supplier to purchase raw materials in advance, thereby reducing lead times and, possibly, costs.

Necessity for Flexible Contract Types

In some situations the buyer's needs are unique and do not fit the supplier's "boilerplate" contract. Negotiating allows for the development of a flexible contract or agreements. For example, the buyer may require a unique delivery schedule, have irregular volume, require services that the supplier does not ordinarily provide, or be unwilling to make the type of commitment to which the supplier is accustomed. All of the terms and conditions should be carefully detailed during the negotiation so that the contract drawn up by the legal department matches the verbal agreement made during the negotiation. It is important that counsel is comfortable with the document from a legal standpoint and that purchasing is comfortable from a commercial standpoint.

Lack of Firm Product Specifications

Vague product specifications may result in inconsistent interpretations by the buyer and seller that require negotiated settlements. For instance, vague quality specifications may mean that the supplier interprets the specifications one way and conducts testing and quality control accordingly, while the buyer's quality control may interpret the specifications completely differently. As the supplier's product moves through incoming quality control, discrepancies will arise. If measuring and testing devices used by the two companies are different, the result could be rejections during incoming quality inspection.

Single-Source Strategies

In situations where there is no active competition, the supplier may have low motivation to provide customer service or engineering assistance. The buyer has limited opportunities even in a negotiation unless there is some other compelling reason for the supplier to desire the contract. On the other hand, the buyer may desire a single source because of strategic reasons, good pricing resulting from consolidating the volume requirements, transportation advantages, or technical expertise the supplier possesses. Several tactics are available in a single-source situation:

1.   The buyer can drop the product line if he or she fails to resolve pricing or other issues with the supplier during the negotiations.

2.   The buyer can consider producing the product in-house.

3.   The buyer can appeal to the supplier's sense of partnership.

4.   The buyer can determine how well the business is going for the supplier and whether the supplier has inventory or cash flow problems. If the answer is yes, this could give the buyer some leverage.

Obviously, negotiation is an ever-present activity in the supply area and negotiation skills are a fundamental requirement for any purchaser. Therefore, experience in negotiation is valuable and proper planning is a must.

Planning a negotiation properly is also a seller's preoccupation, as the following excerpt shows:

> Lanny Weaver, vice president of sales and marketing at American Mirrex Corporation, says, "We want all our sales people to use a negotiations planning process. They need to identify both our objectives and the customer's objectives in the negotiations. Then they need to understand relative leverage and limitations on both sides, and to think through some creative alternatives. We also want them to consider both our risks and the customer's risks in failed negotiations. And, finally, they need to be aware of all the background people on both sides who are impacting the negotiation."[1]

The next section provides a framework of chronological steps in the process of negotiation, which can be used both internally and externally.

PLANNING THE NEGOTIATION PROCESS

The negotiation process may be broken down into eight chronological steps:

---

[1]Robert E. Keller, "Preparing for Negotiations: The Sales Side" *NAPM Insights*, September 1992, p. 16.

1. Basic groundwork.
2. Developing objectives for the written negotiation plan.
3. Gaining support.
4. Getting the team together
5. Live action.
6. Verbal agreement.
7. Written agreement.
8. Making it work.

Together these eight steps form a whole, with each step making its own unique contribution. During steps 1 through 5, it is possible to retreat to earlier steps or advance to later ones.

Step 1: Basic Groundwork

Successful negotiations require careful preparation, a full understanding of the relevant facts about the situation, and a good feel for the desired end result and how this might be accomplished. A technically perfect negotiation will be meaningless if the negotiator discovers at the end that he or she negotiated for the wrong thing. Therefore, the preparation step is critical to achieving success.

The preparation step can be divided into several smaller activities:

1. Setting ballpark negotiation objectives.
2. Gathering relevant information.
3. Generating alternatives.
4. Evaluating and selecting alternatives.
5. Reconfirming that negotiation is the best alternative.

Setting Negotiation Objectives

The first item in the negotiation process should be setting the ballpark objectives. Is the key goal higher quality, an accelerated delivery schedule, a lower price, the ability to get out of a long-term contract easily, or something else? It is critical to think about what is really important in the desired end result. These preliminary objectives should not be overly specific before the relevant facts have been gathered. Thinking about the key drivers of the negotiation allows the data gathering to be focused and establishes a feel for the ways in which the objective might be accomplished.

Some of the objectives that a purchaser might desire are:

- Internal agreement to test a new product, service, or supplier.
- Quality improvement.
- Fair and reasonable price.
- Timely performance.
- Meeting the minimum essential needs of the organization.
- Control over contract performance.

Internal Agreement to Test a New Product, Service, or Supplier.
The objective may be persuading a requisitioning department, unit, or person to try an alternative product, service, or supplier. Understanding what the benefits and disadvantages are likely to be from the user's perspective is fundamental to good preparation. Will the user incur extra costs, lose time or set-up, or risk quality or customer satisfaction if the test proceeds? How can the potential disadvantages be minimized? How far is the supplier willing to go to assure success in the test and recompense if failure occurs? Is it possible to visit other sites where the new product or service or supplier is currently performing? Is it possible to talk to other users and obtain their opinions? Are the potential benefits strategic or operational? How open to trying out and experimenting with new ideas are the people that need to be persuaded? How does this fit their personal goals and aspirations?

Quality Improvement.   A significant quality improvement may have substantial strategic and operational benefits. Getting internal agreement to pursue this and willingness to provide additional resources for training, testing, and supplier evaluation has to be a first step. Getting supplier agreement to pursue this vigorously requires identification of potential benefits to the supplier. Similarly, the associated costs and risks have to be identified and dealt with equitably. If the only customer to benefit from the quality improvement is the initiating purchaser, the situation is drastically different from the one in which the supplier can use the quality capability to expand market share.

Fair and Reasonable Price.   Because cost control is a major concern in all purchasing departments, it will probably rank high on the list of objectives. Total cost of ownership, life cycle costing, activity-based costing, and other cost modeling techniques and concepts may be employed to support the negotiator's cost-cutting goals. The negotiator must never lose sight of other critical issues such as good quality, on-time delivery, and supply assurance. Likewise, the supplier and the internal customer must be able to meet the cost and price objectives without losing money on the deal or feeling uncomfortable about the outcome. The price sought should be fair and reasonable to all parties because even though the buyer has the upper hand now, he or she may be at a disadvantage later and seek the same standards of fairness and reasonableness from the supplier.

Timely Performance. Performance issues, such as shipment according to the production plan, are important negotiating points. The supplier should know that its performance will be tracked and measured against the agreement and other suppliers. Both early and late shipments are undesirable. Early shipments lead to unnecessary inventory and carrying costs and late shipments cause expensive down time. On the buyer's side, accurate production schedules and timely notice of needs will enable the supplier to run its operations smoothly and efficiently. Similarly, persuading internal customers to set production or use schedules reflecting the supplier's capability may be required.

Meeting the Minimum Essential Needs of the Organization. At times, the main concern of the negotiator may be simply meeting the essential needs of the buying organization. If shortages exist for commodities and components and the market is tight, the buyer will have to secure an adequate supply from an acceptable supplier while operating from a weak bargaining position. If the volume is attractive for a long-term commitment and the buyer is a valued customer, the buyer may be able to negotiate satisfactorily. The buyer may gain some leverage by assuring the supplier continued business even after the shortage is over. If shortages cannot be overcome, the internal customer may have to be persuaded to adjust to reality or assist in looking for substitutes to alleviate the situation.

Control over Contract Performance. Performance issues and the responsibilities of all parties should be clearly defined in the contract. In the preparation stage, the buyer should list all issues that can cause performance problems and review the steps to be taken to resolve performance problems should they occur. Possible issues are inability to ship on time or in proper quantities, quality problems, price escalation, buyer's inability to purchase the contract volume, and cancellation clauses and costs. In internal procedures, getting agreement on purchasing's responsibilities and roles with users, engineering, or others, falls into a similar class of negotiation.

Gathering Relevant Information and Defining Needs

Conceptually, the framework for gathering relevant information is rather simple. Short- and long-term information needs to be gathered for our side, and also about the other side. Furthermore, information needs to be gathered about the strategic and operational implications at an economic and psychological level. See Figure 6-1 for an overall summary of data requirements.

The psychological level requires answers to questions like, What kind of person am I going to be negotiating with? What are her or his short- or long-term needs and how can I recognize them in planning this negotiation?

The short-term information can be reasonably easy to gather. It is more difficult to gather information for the long-term perspective, but it is vital nevertheless. Purchasing research has as one of its responsibilities the forecast of future organizational requirements and future market conditions for key requirements. Therefore, the impression should not be given that negotiation requires exclusive information not normally of concern in other supply activities such as strategy development and planning.

Obtaining information is a time-consuming process. Much of what is readily available may not be relevant and much of what is critical to the decision may be difficult to obtain or unclear. For example, what are interest, inflation, and unemployment rates going to be like for the next five years? What new technology is coming along that may make this equipment obsolete?

FIGURE 6-1
Gathering Data for Negotiation

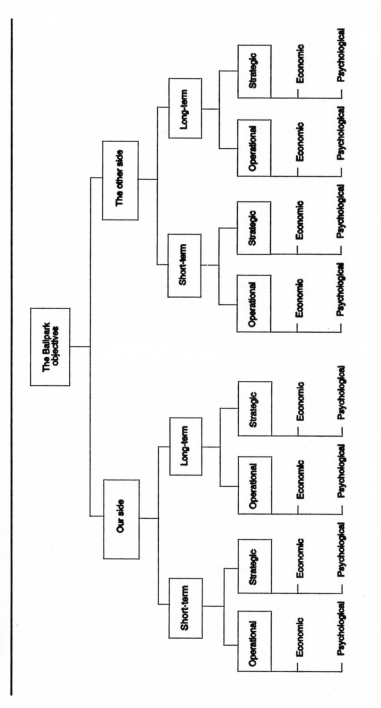

Generating Alternatives

Once the needs of each side are defined, it is useful to develop a few ideas about how to achieve what is desired. Brainstorming is a good way to come up with as many alternatives as possible in the hope that at least one winner will be on the list. After generating a long list of alternatives do some analysis and narrow the list down to two or three alternatives with a high probability of success.

Additional information may be needed once the list is narrowed. It is easier to gather information about specific ideas than it is in the abstract. Having more than one option to work with allows for contrast and comparison, and as alternatives are examined one by one, other possibilities may arise. As information becomes available on one aspect, the potential for a different twist or another direction may become apparent. Remaining open to other alternatives throughout this preparation stage helps to ensure the selection of the best alternative possible.

When examining alternatives it may be helpful to:

1. Ask, If this idea turns out to be unworkable, what others are left?
2. Make sure there are at least three to six reasonable ideas, and don't stop until you have them.
3. Ask at least one other person and possibly more for ideas.
4. Have a brainstorming session.
5. Change the conditions of the short term and the long term and see how it affects your options.
6. Try to see the same situation from the other side's perspective and look for alternatives.
7. Change your assumptions and see if it changes the alternatives.

When narrowing the alternatives, possible reasons for discarding an option are: no relevance to the situation; low probability of success; too costly; not practical under the circumstances; too difficult under the circumstances; would create too many bad aftereffects or consequences, even if successful; too risky; and morally, legally, environmentally, socially, culturally, or politically unacceptable.

Evaluating and Selecting Alternatives

After selecting one alternative to pursue, a review must make sure the required information is available and complete. Check the information by asking:

1. Have I the necessary information for both the short term and the long term?
2. Can I justify the alternative on a qualitative basis, a quantitative basis, or both?
3. Is there anyone inside or outside the organization who knows more about this than I do? If so, in that person's eyes is this information relevant, and should I get more?
4. If I had the time, would I try to collect more information? If so, what and why?
5. If I had to pass this situation on to someone else who knows nothing about it at the start, would he or she be able to take over with the information I have?

A large part of information gathering and analysis involves detailed cost/price analysis techniques such as value analysis, learning curve, and total cost of ownership (discussed in Chapter 5).

Reconfirming That Negotiation Is the Best Alternative

After identifying the alternatives to be pursued, it is wise to reconfirm that negotiation is the best alternative under the circumstances. The two prime arguments for selecting negotiation have to be (1) the results expected using negotiation are superior to those obtainable by other means and (2) the extra cost of negotiation is worth the extra benefits expected.

Step 2: Developing Objectives and the Written Negotiation Plan

After determining alternatives and gathering information, the negotiator is ready to start developing objectives and strategies in an effort to develop a written negotiation plan. The purpose of this second step is to determine what results can be achieved reasonably and how to achieve them. This involves the following activities:

1. Analyzing the position of strength of both sides.
2. Setting detailed objectives.
3. Developing the negotiation plan.
4. Selecting the team.
5. Deciding on timing.
6. Picking a negotiation site.
7. Choosing tactics.
8. Writing the negotiation plan.

The sequence of events is important because the points giving one party strength over the other party will help that party establish its objectives and outline for the negotiation strategy.

Analyzing Positions of Strength

The desire for agreement, the perceived certainty of getting the agreement, and the commitment or intention to perform after agreement depend on the relative strengths and weaknesses of each side at the time of the negotiation. One critical element in preparing for a negotiation is assessing the positions of strength of both or all parties. Performing this task helps establish future negotiation points, avoids setting unrealistic expectations, and may reveal ideas about possible strategies.

Analyzing the factors that give either party a position of strength should be done in light of the key ballpark objectives of the negotiation. Knowing which objective is the most important one will dictate the position of strength of either party entering the negotiation. The key objective of one party may not be the key objective of the other. Figures 6–2 and 6–3 give 40 potential points of strength for purchasers and suppliers.

Purchasers often underestimate their own position of strength. Even though this list is extensive, it is by no means complete. In most purchasing negotiations, the purchaser should have at least 12 to 24 points of strength with at least four major ones. Clearly, every individual situation is unique, and in this design or planning stage of the

FIGURE 6-2
Forty Potential Points of Strength for a Purchaser

1. Pays on time
2. Financially strong
3. Excellent reputation
4. Significant market share
5. Large volume requirement
6. Substantially growing future requirements
7. Steady use or counterseasonal requirement
8. Geographical proximity to supplier
9. Strong technical expertise
10. Can add additional products or services to the requirement
11. Can influence other purchasers
12. Long history of successful supplier relations
13. Innovative reputation
14. Global strength
15. Only or large buyer in the market
16. Larger organization than the supplier
17. Accurate forecasts
18. Stands behind commitments
19. Ability to make the requirement in-house
20. Other suitable suppliers are available
21. Substitutes are available
22. Buyer's market
23. Strong currency
24. Strong organizational support for purchasing
25. Knowledgeable negotiators
26. Skilled negotiators at the technical, economic, and human levels
27. Aware of supplier weaknesses
28. Well prepared
29. Willing to go long term
30. Can and will help suppliers to improve
31. Organizational values and goals are congruent with supplier's
32. Flexible and willing to adjust
33. Ethical, honest, and trustworthy
34. Good government relations
35. Good public image
36. Strong brands or patents; has a sustainable competitive advantage
37. The contract under negotiation can help reduce the supplier's variable costs or contribute to sunk fixed costs
38. Tolerant of mistakes if repetition will be avoided
39. Potential for favorable public relations or image impact
40. Sufficient time for proper negotiation

negotiation, negotiators should carefully assess and record their own position of strength as well as that of the supplier.

The purchaser's individual points of strength can be mirrored on the supplier side; some, such as a solid reputation and sufficient time, are exactly the same. Figure 6–3 lists the potential strengths a supplier may have.

Analyzing the position of strength of each party is important for two reasons. First, the analysis is a prerequisite to setting objectives, and second, the individual points in the position of strength identify what each side will actually say during the negotiation. This step helps prepare the negotiation strategy and plan for the face-to-face negotiation or live action. It also helps identify the key arguments that the other side will use in their strategy and allows time to prepare for counterarguments and ways of dealing with the other side.

After carefully listing everything that gives each side strength, it is necessary to ask, Whose position is stronger? This is a judgment call. By determining the relative strength of each side the buyer can set more realistic objectives. It is also useful to ask, Which points are the stronger ones for each side? This allows the negotiators to stress and elaborate their key strong points and be prepared for the arguments from the other side.

In a normal buyer-supplier situation, the buyer seeks out strong, reliable, capable suppliers who can be counted on to provide what is needed. These same suppliers will also have relatively strong positions because of their superiority as suppliers. So, although the buyer may want to have a strong upper hand in a negotiation, it is natural that the most desirable suppliers will also operate from strong positions. Negotiations with strong suppliers may be difficult, but this situation is preferable to negotiating with weak, unreliable, incompetent suppliers. That is why purchasing may be defined as the fight for good suppliers. The reputation of the purchaser can be assessed by the quality of the suppliers the purchaser can attract.

FIGURE 6–3
Forty Potential Points of Strength for a Supplier

1. Superior quality; ISO 9000 certification
2. Superior quantity capability
3. Superior delivery
4. Superior service
5. Superior price
6. Superior terms and conditions
7. Superior warranties
8. Superior flexibility as to specifications, delivery, quantity, timing, terms and conditions, packaging, and special requirements
9. Superior research and development
10. Financially strong
11. Excellent reputation
12. Significant market share
13. Future growth assured
14. Strong brands or patents
15. Long history of successful relations with this customer
16. Reputation as innovative
17. Reputation as environmentally superior
18. Global strength
19. Close proximity to buyer
20. Only supplier available
21. Larger organization than the buyer
22. Stands behind commitments
23. Has superior suppliers
24. Seller's market or market shortage
25. Government regulations or requirements favor supplier
26. No substitutes available
27. JIT and EDI capability
28. Knowledgeable and skilled negotiators at the technological, economic, and psychological levels
29. Aware of the purchaser's weaknesses—knows more about the purchaser's organization than most of its own employees
30. Well prepared
31. Can and will help purchaser to improve
32. Organizational values congruent with purchaser's
33. Ethical, honest, and trustworthy
34. Good government relations
35. Good public image
36. Tolerant of purchaser errors if repetition will be avoided
37. Potential for favorable public relations or image impact
38. Sufficient time for proper negotiation
39. Close to capacity
40. Other purchasers anxious to buy

Setting Objectives

Objectives can be set more fairly if an honest attempt has been made to look at the relative strengths of each party. This avoids unreasonable frustration with overly ambitious expectations on the one hand or too easily achieved objectives on the other. The term "what the traffic will bear" reflects some of the common-sense wisdom behind this approach. The circumstances determine what is reasonable. For example, when a university student asked a security guard why some person would take an old plastic shopping bag off her bike's saddle in a rainstorm, the security guard said, "In a rainstorm, that bag was worth a lot."

The kind of objectives being set at this stage are "come home with" objectives. Under the circumstances, what are reasonable results to "come home with" after the negotiation? These "come home with" objectives are different from tactical objectives communicated as part of the bargaining process. For one reason or another, a party may wish to start the bargaining process with extreme demands with the intention of moderating later on. The "come home with" objectives represent reasonable results to be achieved by a negotiator who is not depending upon luck to achieve them and who is well prepared.

A standard way to set objectives is to establish a range within which a negotiator is willing to settle. For example, in the purchase of a home, a buyer may expect to pay somewhere between $220,000 and $250,000 as the maximum affordable price. In diagram form this can be represented as follows:

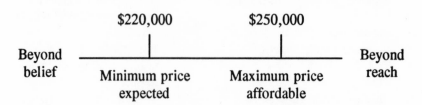

The seller of the home, unaware of the purchaser's expectations and ability to pay, may set a range of prices within which the seller might be willing to sell. Now three possibilities present themselves:

1.   The seller and purchaser overlap.

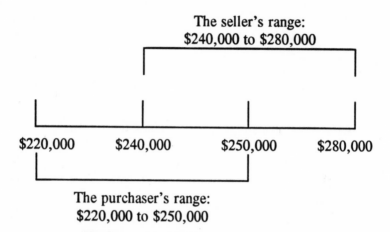

The seller's range:
$240,000 to $280,000

$220,000 $240,000 $250,000 $280,000

The purchaser's range:
$220,000 to $250,000

Presumably, barring the willingness of either side to move the range, the overlap zone between $240,000 and $250,000 becomes the area within which the two sides might be able to settle.

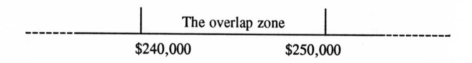

The overlap zone

$240,000 $250,000

2.   The seller and purchaser do not overlap.

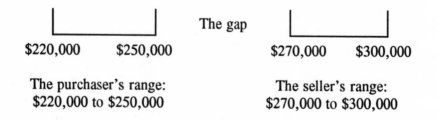

The gap

$220,000 $250,000 $270,000 $300,000

The purchaser's range: The seller's range:
$220,000 to $250,000 $270,000 to $300,000

In this case, there is a gap between the upper limit of the buyer and the lower limit of the seller. This gap of $20,000 now means that if a deal is to be reached, one or both parties must be willing to adjust the range to achieve a zone or point of overlap. In this second scenario, the likelihood of a successful deal is substantially less than in the first case.

3.   Other possibilities

The other possibilities that exist include the following:

a.   Both sides meet at one point only. For example, the upper limit of the purchaser coincides with the lower limit of the seller.

Purchaser's range	Seller's range

$220,000 $250,000 $280,000

Provided that both sides can negotiate their way to the $250,000 point, settlement is still possible here.

b.   The seller's range could, theoretically, be below the purchaser's range. For example, the seller may be willing to sell for $180,000 to $200,000. This kind of situation could arise if either side or both have not done sufficient preparation or if other factors are seen by one side as more important than price. For example, the seller may want cash within 48 hours, or the purchaser may want possession within one week. This is called the surprise side of negotiation, and either side may be cautious about revealing its range. The possibility that afterward one or both sides may feel they could have done better is fairly high in this type of situation.

c.   The seller's range is so much higher that no amount of negotiation will help. For example, if the seller's range starts at $500,000 when the purchaser's ability to pay does not extend beyond $250,000, the negotiation may be meaningless.

In conclusion, preparation for the negotiation should disclose whether the possibility of a deal exists and whether the negotiation falls into an overlap zone or a gap situation. Substantial gaps may be difficult to bridge, and recognition that the negotiation may not be successful may force an early search for alternatives. The ability to close the gap with the help of position of strength points allows the negotiators to achieve agreement. For example, if the purchaser's ability to pay is limited by the mortgage company's willingness to provide funds, the seller may offer a second or third mortgage or a mortgage at a lower rate without reducing the price. The negotiation plan provides options for the negotiator to deal with any of the possible starting points discussed.

Developing the Negotiation Plan
Once the negotiator knows what the objectives are, he or she must figure out how to achieve them. To establish the negotiation plan, the negotiator must answer the questions of what, why, who, when, where, and how.

What and Why. Although the what and why should be pretty well established at this point, it is a good idea to review what is being negotiated and why. Any lingering doubts about either point should be resolved before proceeding.

Who. Careful thought should be given to who will be doing the actual face-to-face negotiation. Some characteristics of good negotiators are:

Ability to gain respect	Patience/Tolerance
Ability to listen for ambiguity	Persistence
Analytical ability	Personal integrity
Competitiveness	Planning ability
Control	Problem-solving ability
Cross-cultural strengths	Self-esteem
Decisiveness	Sense of timing
Ethical behavior	Tact
Insight	Verbal clarity
Intelligence	

Should an individual conduct the negotiation or should a team? If an individual is to be selected, should he or she be a recognized expert, a skillful negotiator, a person of formal authority in the organization, or someone of prestige outside the organization? If a team is to negotiate, who will compose the team and what will be the role of each individual? A team is often used when a contract is complex or long-term and one person will not have sufficient knowledge to cover all issues, or when the purchase is strategic. The team may consist of top management, purchasing, engineering, quality, operations, finance, and marketing. Any number of people can be moved into and out of the negotiation depending on the length and complexity of the negotiation.

One means of deciding whom to select for a negotiation team may be to consider the individuals who think the organization has the strongest position. After all, perception influences behavior. If an individual believes he or she has a strong negotiating position, he or she may negotiate better.

When. Timing is quite often a crucial element in a negotiation because the party under the greatest time pressure is usually at a disadvantage. It may be possible to structure the negotiation to gain a time advantage. For example, if a supplier is embarking on a new production process and a buyer wants to structure the most advantageous deal, the timing of the negotiation might affect the final price. Is it better to approach the supplier before the equipment is installed and running? Or is it better to wait until just before the equipment is installed, when the supplier may be feeling nervous about the investment and the possible costs of the initial production runs? Or is it better to wait until after the equipment is installed? And if so, how long after installation? If the initial production runs go smoothly, the purchaser may be able to drive a harder bargain. If the initial runs are fraught with quality problems, delays, and extra costs, then the buyer may be able to offer technical assistance in return for a good price.

A negotiator should have time goals about when to conduct the negotiation and how long a contract term should be.

Where. The location of the negotiation should be carefully chosen. Generally, there are three choices: on Party A's turf, on Party B's turf, or on neutral ground. It is always best to conduct negotiations on one's own turf because one is then essentially in control. At home, one's confidence and comfort levels are higher, data and support are accessible, stress is lowest, and travel fatigue is eliminated.

However, it is not always possible or advisable to negotiate on home ground. If the buyer is the initiator, he or she may have to go to the supplier to make a proposal. A neutral location may be desirable to remove the perception of advantage for either. If a neutral site is chosen, it may be a site closer to Party A than Party B, or vice versa, or some midpoint location. A neutral site might be a restaurant, a trade show, a conference, an industry meeting, or a golf course.

The negotiation plan requires deciding how much information will be provided ahead of time. Suppose a buyer wishes to feel out a supplier about a possible venture. If the buyer tells the supplier too much up front, the supplier may reject the idea out of hand. If the buyer does not provide enough information, the supplier may not even be interested in meeting. Sometimes, a "chance" meeting at a trade show or some other common ground may be the optimal time and place to pitch an idea.

How. How one negotiates or the tactics one employs is driven by the analysis of positions of strength. A tight market, limited sourcing, and the urgency of the deal are examples of issues that have a bearing on the tactics used. Some of the tactical options are:

1. Start on principles—if they buy it in principle, the rest will be easy.
2. Horse trade, start high, and haggle.
3. Kill them with kindness—ply them with food and drink and friendliness.
4. Delay—find ways of slowing down your side until they can't stand it any longer.
5. Plead—ask for the milk of human kindness.
6. Imply future consequences—"You may have me over the barrel this time, but wait until next time."
7. Use a high-level approach—get their boss to agree.
8. Use experts—blind them with qualifications.
9. Agree on the little things first—if we can agree on the little things first, this will make it easier to deal with the tougher aspects later.

10. Agree on the big things first—if we agree on the big things first, the little things will take care of themselves.
11. Seek precedent—use precedent to push for bigger bites later on.
12. Play one side against the other—use other options to put on pressure (if you don't agree, someone else will).
13. Use a team—use a number of people to overwhelm the other side.
14. Seek win-win options—find ways of achieving mutual benefit for both sides.
15. Stick, stick, stick—clarify demands and detail why they are reasonable and insist they cannot be changed.
16. Feign ignorance and ask for help—"I'm no expert in this business, so please help me resolve this."
17. Shift emphasis—when you really want a good price, insist on superior quality.
18. Threaten deadlock—"It looks to me as if a deal is not possible."
19. Walk out—physically remove your side.
20. Switch players—get fresh troops in.
21. Build roadblocks—"Our personnel manager will not be able to go for this."
22. Take it or leave it—"This is your last chance."
23. Use any combination of the above or others.

In his book entitled *Give and Take: The Complete Guide to Negotiating Strategies and Tactics*, Chester L. Karrass[2] states that the choice of negotiation tactics is limited by strategy. One should not win a short-term gain by compromising a long-range goal. Karrass goes on to say that one must define all issues, problems, and goals, tie them together, and come up with tactics that will satisfy them.

Some additional negotiation tactics include the following:

Agenda—Establishing an agenda is a tactic whereby either party informs the other of issues it considers important. Either party can

---

[2]Chester L. Karrass, *Give and Take: The Complete Guide to Negotiating Strategies and Tactics*. New York: Thomas I. Crowell, 1974.

then play up or play down specific points on the agenda. This is a method for indicating significant issues worth discussing.

Deadline—Establishing a deadline can put pressure on both parties and may result in a settlement that is not satisfactory to either party. Deadlines are not always met. Nevertheless, negotiations can continue, and a satisfactory settlement can be reached even after a deadline has passed.

Concessions—Making appropriate concessions can produce significant gains. Concessions need not be made "one for one" or be of equal value. A large concession by an opponent need be matched only by a less significant one or perhaps by none at all. Concessions should be used carefully, however, because they will set the tone and momentum for the entire negotiation.

Missing person—The missing person tactic is the deliberate absence of the person with final authority from the negotiating table. It provides the negotiator with more time and sometimes a way out of a tight situation. Using this tactic, the negotiator tells the opponent that he or she must return to the home office with information in order to gain the final decision.

"The bogey"—In this tactic one tells the seller that he or she likes the product, but that $X is all that can be paid.

"Take it or leave it"—This tactic lets the other party know that one is firm on an issue and will not move. This can deadlock the negotiations and cause them to be delayed for an extended period. Prior to using this tactic, consideration should be given to the consequences; usually, it should be the last resort.

Best and final offer—Telling your opponent that this is the best and final offer (also known as the "doorknob" price) is a very risky negotiating tactic that the buyer must thoroughly believe in or else risk losing the deal. In his book *Give and Take*, Chester Karrass[3] explains the importance of how the best and final offer is presented. Different wording will convey different meanings. Consider the following statements:

- "This is my final offer for the product."
- "This is my final offer considering all factors."
- "If you don't accept my offer in 3 days, call me."
- "If you don't accept my offer in 3 days, consider the deal dead."
- "If you don't accept in 3 days, I will place my order with another supplier."
- "If you don't accept the offer in 3 days, we will still be friends."

The best and final offer can improve or hinder bargaining power. The manner in which it is stated will convey commitment and, most important, will leave room for possible retreat or the ability to continue talking if necessary.

If the buyer is presented with a best and final offer in a negotiation, he or she should listen, try to determine if it is a bluff, and let the supplier know that it may cause a deadlock. If the buyer can see the tactic coming, it may be a good move to "beat the supplier to it."

The Written Negotiation Plan
The final product of the second step in the negotiation process should be a written negotiation plan. At a minimum, this plan should identify:

1. A brief overview statement.
2. Why the negotiation is required, what is to be negotiated, the key objectives, and with whom the negotiation will be done.
3. The statements of position of strength of both parties.
4. Whose internal support needs to be gained, and how and when it will be obtained.

---

[3]Karrass, p. 153.

5. Who will be on the negotiation team, why, and how the duties will be split.
6. The negotiation plan and tactics: where, when, how, and for how long.

Preparing this negotiation plan may be expensive and time-consuming, but it is essential to successful negotiation. The written plan forces discipline, permits communication and takes the luck out of negotiation to the maximum extent possible.

Step 3: Gaining Support

The prime purpose of gaining support is to ensure that the written negotiation plan is acceptable internally. Before approaching the other party in a negotiation it is important to have full internal support and to be in conformance with organizational strategies, initiatives, practices, policies, and procedures. Two questions must be answered: (1) Whose support is needed? and (2) What type of support or approval must be gained? For example, if purchasing is negotiating something with accounting, then the internal supporters are the other people in purchasing and on the other negotiation side are the people in accounting.

Deciding whose support is needed requires knowledge of how the organization works and who has formal and informal power. If the support of several people is needed, consideration must be given to how these individuals work together regularly. Who is approached first may affect the degree of support attainable from the other targeted individuals.

During the preparation and planning stages some internal support may have been gained from the people providing information, data, or alternatives. Building support gradually in the initial stages helps ensure ultimate support for the complete plan. An inability to get the needed support will force review and possibly revision of the objectives and negotiation plan.

Step 4: Getting the Team Together

If only one person is negotiating, then this step should review the work done so far. If a team will be negotiating, then meetings should be held to discuss goals, plan strategies, and review strengths and weaknesses of both sides. A team leader should be selected to guide the team during the negotiations and chair any prenegotiation meetings. The leader should be the main contact point between the buyer and seller and should set up times and locations for the negotiating sessions.

Each person on the team should have a clearly defined role he or she is expected to play, and he or she should understand and agree to this role. Nothing is worse than arguing and disagreeing among team members in a negotiation session. It is critical to present a common front.

Step 5: Live Action

Live action is the execution of the strategy. Although the actual negotiation could take place via telephone, mail, facsimile, or electronic mail, the face-to-face approach is preferred for most important negotiations.

The live action may be broken into five phases: the presentation, clarification, exploration, proposal, and hard bargaining phases. Although it may be difficult to delineate these phases sequentially because of overlap and backtracking, the following activities take place with each phase.

Presentation Phase
During the presentation phase, both sides make introductions and opening remarks and then present their general positions. This phase is often characterized by rhetoric on both sides designed to warm or soften up the other party about negotiation priorities or conditions. Team members have a chance to say a few words and get comfortable. The statement of general positions may be lengthy or brief, specific or rather philosophical, but should generally be heard as opening gambits.

Clarification Phase
The clarification phase is the time for both sides to make sure they understand the details of each other's position. What do you mean by "we stand behind our product?" By careful listening and probing, the true meaning of the other side's statements may be discerned.

Exploration Phase
The purpose of the exploration phase is to establish the boundaries and priorities of the negotiation. It is necessary to delineate the points of give and take. Agreement may be reached on some minor items, but nothing of real significance is normally given up at this stage. Rather, the time should be taken to test statements for their validity, to clarify where bluffs might lie, and to determine where it is possible to give and take. This phase is characterized by a successive unfolding of each other's position. The difficulty is deciding what to reveal, when to reveal it, and how to respond to revelations by the other party.

Proposal Phase
After exploration, either or both parties will put forth a more detailed proposal, which may contain significant concessions from the starting position. Often the proposal will contain statements like, "We are willing to be flexible on delivery and price, if you will commit to a long-term contract and agree to meet our quality specifications." This phase is characterized by proposals and counterproposals.

Teams may take frequent breaks from the negotiation to discuss particular points of a package or to prepare a counterproposal. Sometimes negotiations may be discontinued to give one or both parties an opportunity to confer with other people in the organization or to gather additional information or consider alternatives.

Hard Bargaining Phase
Hard bargaining may begin near the end of the negotiation, when time pressure may be severe for one or both parties. Detailed trade-offs have to be agreed to one by one. This phase is characterized by time pressures, emotional vulnerability, and the reality of the situation. Controlling emotions can be particularly difficult after lengthy negotiations if one or both parties are tired, tense, and anxious to settle well. If the actual negotiation is not going as well as the planned strategy

it may take considerable self-restraint to conclude on the best note possible.

Step 6: Verbal Agreement

The last step in the live action is the point at which one party says "I think we have a deal" and the other party agrees. A good negotiator will compare the verbal agreement to the key objectives of the original plan and determine if all objectives have been met satisfactorily. Any uncertainty must be resolved before agreeing to the verbal deal.

Once both sides concur that an agreement has been reached, it is important to review the agreement point by point until all points are mutually acceptable. This seemingly small step may prevent problems later.

The verbal agreement should then be put into writing, in a letter, memo, contract, draft, procedure, notice, or purchase order. A well-supported and reviewed verbal agreement forms the basis of the written deal. A letter of intent is a good interim document if there will be a significant time lag between the verbal agreement and the contract preparation. Both parties may base their subsequent planning and actions on the letter of intent. (See Chapter 7 for more about letters of intent.)

The negotiation objectives and the process of achieving those objectives are closely related. The way in which the negotiation is conducted and the way in which the verbal agreement is reached may seriously affect the future behavior of both parties. There is a huge difference between ending a negotiation by saying, "Well, I guess I have no choice but to accept since I really have no alternatives," versus, "I look forward to our working together on this project." If intention to perform is the ultimate goal, the level of commitment may be different for the speaker in each of the previous statements.

Step 7: Written Agreement

The written agreement should confirm the verbal agreement. Lawyers should be involved in the contract preparation if there are legal implications. Ideally, nothing in the written agreement will invalidate the verbal agreement. In reality, the written agreement may result in another face-to-face negotiation all over again. One or both sides may find agreement is no longer possible once the written version is examined. If the contract includes legal or technical expressions, or covers seemingly irrelevant points, the written agreement may not resemble the verbal.

The focus must be on ending up with a written document to which both parties are committed. Even if negotiations must be reopened, this may be preferable to living with an unsatisfactory agreement.

Documentation of Negotiations

Because of personnel turnover and the frailties of human memory, accurate documentation of negotiations is essential. Documentation should include the following:

- *Subject*—A memorandum giving an overview of the negotiation, including supplier's name and location, contract number, and a description of the requirements to be purchased.

- *Introductory summary*—The type of contract and the type of negotiation action involved, plus comparative figures of the supplier's proposal, the buyer's objectives, and the negotiated results. Was the written negotiation plan successful?

- *Particulars*—The details of the product or service to be purchased and who is involved in the procurement.

- *Procurement situation*—The factors in the procurement situation that affect the reasonableness of the final price.

- *Negotiation summary*—The supplier's contract pricing proposal, the buyer's negotiation objective, and the negotiation results, tabulated in parallel form.

Step 8: Making It Work

Just because both parties agree to the written contract does not mean the job is done. Living up to the agreement is the real test of the success or failure of a negotiation. It is necessary to make sure both sides live up to the agreement. The negotiator should act as a watchdog in his or her own organization first and of the other side second. In some organizations contract administrators see to the management and execution of the agreement. Problems must be resolved within the framework and intent of the agreement, even if no specific remedy is provided in the written agreement.

Since current behavior affects future expectations and behavior, the way in which conflict is resolved will significantly affect any future dealings. Preserving and building trust during an existing contract may make future negotiations easier. For example, if a payment is going to be late, does the buyer alert the supplier ahead of time, or does he or she ignore the phone calls of the supplier? Seeing a contract through in every detail requires consultation, communication, and cooperation. It is inappropriate to avoid an issue such as late payment on the grounds that accounting handles payment.

Negotiations are especially difficult when conducted to resolve a problem or a soured deal. This situation may be the best test of the skill and ability of a negotiator. How problems are resolved reflects on the completeness and thoroughness of the deal. Changes in contracts, including early termination, if not provided for in the original deal, need to be negotiated. If earlier performance was unsatisfactory, negotiations will be more difficult than if good intentions and conformance to expectations were the pattern.

Negotiation has been called both an art and a science. While some individuals may seem more naturally adept at negotiation, it is a skill that can be learned. Experience and practice are the best teachers. Both formal and informal opportunities to negotiate abound. One step to being a better negotiator is to take a few minutes to evaluate performance after any negotiation, large or small. It is useful to ask four questions:

1. How did I do?
2. Why did the negotiation go the way it did?
3. What have I learned?
4. What should I try to do better the next time?

KEY POINTS

1. Negotiation is a process whereby two or more parties try to reach agreement.

2. Proper planning is the key to negotiation success.

3. Purchasers need to be able to negotiate effectively within their own organizations as well as with suppliers.

4. The fundamental purpose of negotiation is to reach agreement in such a way that both parties will perform as they agreed.

5. There are many reasons why negotiation may be chosen or be necessary. These reasons may be internal or external to the organization.

6. The negotiation process can be divided into eight chronological steps: (1) basic groundwork; (2) developing objectives and the written negotiation plan; (3) gaining support; (4) getting the team together; (5) live action; (6) verbal agreement; (7) written agreement; (8) making it work.

7. Steps 1 and 2 in the negotiation framework form the basis for the written negotiation plan. Steps 3 to 8 are the execution of this plan.

8. Negotiation skills can be learned and improved with experience and continuing self-assessment.

CHAPTER 7

CONTRACT AWARD AND ADMINISTRATION AND ETHICS

This chapter covers the remaining steps in the acquisition process. After the bid analysis and negotiation, the purchase order or contract should be awarded to the supplier offering the best overall value in the short and long term. Follow-up, expediting, receipt, inspection, invoice checking, payment, and record maintenance complete the cycle. In many organizations, disposal is also an issue; although it is extensively covered in Books 3 and 4 of this series, it will be briefly mentioned.

The administration of major contracts is a special area. In both the public and private sectors, contract administrators attempt to bring major contracts in on time and within budget. Therefore, a special section is included, highlighting the administration of major contracts.

This chapter concludes with ethical considerations throughout the acquisition process. The legal issues described in the next chapter closely interact with the contract award and administration issues in this chapter. Therefore, both chapters should be cross-referenced carefully to ensure full understanding of acquisition procedures and their legal implications.

PURCHASE ORDER PLACEMENT OR CONTRACT AWARD

Placing a purchase order or awarding a contract should come as the next logical step after bid analysis and negotiation. The supplier selected should be the one that best meets the criteria originally established by the purchaser to meet operational and strategic needs in the short and long term.

There are a variety of ways in which an order can be placed, and the method of procurement chosen will affect the format.

Methods of Procurement

The two main methods of procurement used by purchasing departments are purchase orders and contracts. A third category, bypassing the purchasing department, is user-generated for minor requirements that can be handled by petty cash or credit card, or user acquisition requirements for which the purchasing department is not responsible. An example of the latter may be the acquisition of strategic consulting services handled directly by the chief executive or the board of the organization.

Purchase Orders

The purchase order is still the most common vehicle used to describe the legal commitment between a buyer and seller. It can be a multipart, pre-printed document or it can be an electronic form. In paper form, normally a number of copies are made for use by purchasing (2), requisitioner (1), supplier (1), quality (1), receiving (1), accounting (1), and possibly others. Normally, each purchase order has its own unique number; in large organizations it may carry a code identifying the issuing buyer.

The critical information pertinent to the commercial dealing is normally set out clearly in the first parts of the purchase order: what is being acquired, the amount required, when it is to be delivered and where, what price is to be paid and when, and how it is to be shipped. These primary features are followed by a number of terms and conditions, which may well be printed as standard on each purchase order form. These conditions are intended to protect the purchaser's interests in areas such as the following:

1. Prices fixed for duration of contract and inclusive of all taxes and other applicable duties or levies.
2. Buyer's right of inspection and rejection.
3. Buyer's right to make specification or design changes.
4. Holding the buyer harmless against patent infringement and other legal considerations.
5. Limitations on the right of the supplier to assign the contract to third parties or generate publicity.

6. Rights of the buyer to acquire the necessary goods and services elsewhere if the supplier is unable or unwilling to deliver.
7. The use of commercial arbitration as a contract dispute resolution.
8. That the agreement is entirely covered by the purchase order and no pertinent agreements exist outside the document.

The disagreement between these purchaser-generated terms and conditions and the supplier-generated terms and conditions printed on the standard acknowledgment form creates the "battle of the forms" referred to in the next chapter.

Purchase orders can be configured in a variety of ways.

Blanket Orders

Blanket orders are generally used when the prices quoted are for "any quantity" or are based on "estimated annual requirements." When this method is used, the buying organization places with the supplier a purchase order that identifies the product or service to be purchased and the cost of the product or service in the appropriate unit of measure, along with other applicable terms of the purchase. The buying company then issues material releases or requisitions to identify the quantity to be purchased within a specific period of time against the base (or blanket) purchase order. Material releases are generally repetitive and define requirements on a regular interval (for example, weekly or monthly). Requisitions issued against a blanket order are more likely to address periodic requirements for items required repetitively but not regularly.

Standing Orders

This ordering system is usually arranged through negotiation. It consists of the supplier delivering goods at fixed, agreed-upon prices covering a defined period. This form of ordering is also known as "open-ended ordering" and is often made with rail and motor carriers agreeing to deliver a certain number of cars or trailers per day unless told otherwise.

Purchase Order Draft or Check with Order

In this method, the purchase order document combines the PO with a blank check for payment purposes. The supplier completes the blank check, which is limited to relatively low dollar amounts, and sends it to the bank for processing.

This type of system is also called the Kaiser check system for the company that originated its use. It was an early attempt to improve supplier delivery performance by assuring prompt payment upon shipment. It also reduced paper costs and permitted discounts for early payment.

Credit Cards

Corporate credit cards or procurement cards are issued to authorized members of user departments to charge minor purchases such as MRO supplies within the user department's budget or for items under contract. This practice relieves purchasing of involvement in areas where it can add little or no value. The user also favors this method because of the actual or perceived control over purchases.

Telephone Orders

When the value of the purchase is small and the order is nonrecurring, it may be appropriate to obtain quotation information from suppliers by telephone and also place the PO by telephone.

The same number that is used for the requisition may also be used for the PO. Accounting can then be authorized to pay the requisition. This method of PO placement reduces both the time required to place an order and the associated paperwork.

Consignment

Consignment purchase orders are issued to cover work done on a buyer's material that is physically in the possession of the supplier. For example, a supplier may be contracted to apply paint to a company's in-process product. The product to which the paint will be applied is the consigned material. Purchasing needs to work closely with accounting to ensure that consignment purchases are managed correctly with respect to inventory control.

A different type of consignment buying covers the situation where the seller's goods or equipment are on the buyer's premises while the seller still has ownership. Normally, only when the purchaser withdraws the item from inventory is ownership transferred and the payment process triggered. Thus, the purchaser need not tie up its capital in inventory, but does supply space for storage. Ink suppliers for printing companies often provide consignment inventory; this method is also used in distribution channels as well as retailing.

Hospitals usually have many stocking points within the facility and keep hundreds of supply items on hand. Under a systems contract and consignment inventory arrangement, the supplier monitors usage and inventory levels at each stocking point and replenishes each stocking point from its warehouse. Because the hospital does not have to maintain a central stockroom it saves space and reduces costs.

CONTRACTS

Contracts may be short- or long-term. Annual contracts are still common in many public and private organizations. The decision of whether an annual commitment should be in standard purchase order form or in a contract will depend primarily on past practice, complexity, dollar amount, and its strategic impact. Normally, for repetitive requirements, the current supplier (if its performance was satisfactory) is asked to bid for the next year along with other potential suppliers.

Long-Term Contracts

Long-term contracts are normally for periods greater than one year with no upper limit on time. What may be long-term in one industry or sector may be short-, medium-, or long-term in others. Long-term contracts permit the development of nonstandard and custom procurement processing and sharing arrangements.

Contracts versus Purchase Orders

Legally speaking, purchase contracts and purchase orders are treated equally. In practice, purchasers may prefer to use a contract over a standard purchase order if:

- The contract has a strategic impact on the organization.
- It is standard market practice to use contracts.

- The dollar amount of the purchase is high.
- The length of commitment is for a longer term.
- The complexity of what is being acquired is high.
- The number of items or range of services covered is substantial.
- Board of directors or trustees approval is required for the commitment.
- The requirement is unusual or unique.
- The buyer-supplier relationship goes beyond the standard arms-length commercial practice.

Types of Contracts

A contract may be thought of as a customized, enlarged purchase order. It is normal for all purchase contracts to identify the buyer and the seller; what is to be acquired, how quality will be determined, and what quantity is required (both for the total contract and for individual shipments); what prices will be paid and when; and how, in what form, and when delivery will be made. These first five to eight major sections of the contract normally cover the economic heart of the contract and would normally also be addressed in a letter of intent if that is used instead.

With respect to the price portion of the contract some standard options exist.

Fixed-Price Contracts
Fixed-price contracts are based on a price that will not differ from that agreed upon or understood to apply at the time of ordering. As contract terms lengthen or as complexity of development or performance increases, fixed prices may not be reasonable. Common fixed-price contract types include the following:

> *Firm fixed-price.* This is an agreement for the buyer to pay a specified price to the seller when the latter delivers what is purchased. It requires minimal administration and, from the standpoint of risk avoidance, is the most preferred. In addition to transferring financial risks to the seller, it provides seller incentive to improve efficiency and to hold down costs.

Firm fixed-price with escalation. For contracts involving a long production period and large amounts of money, a seller usually is reluctant to quote firm prices due to inflationary risks. In such cases, a buyer may wish to use an escalator clause, which provides for both upward and downward changes in price as a result of changes in either material or labor costs. Escalation clauses are typically tied to recognized indexes. Such contracts are generally used during periods of economic uncertainty or when the contract is long-term.

Fixed-price with redetermination. Redetermination contracts are used when both future costs and amounts of labor and materials are uncertain. It differs from escalation in that with escalation, the amounts of labor and materials required to complete the contract are known, but the labor rates or the prices of material are unknown. In redetermination, the amounts of labor and materials are initially unknown, but they become known with limited production experience. Typically, the contract starts with a temporary fixed price, which is adjusted after experience is gained and costs are known. Maximum redetermination contracts provide only for adjustment downward from the temporary price. Flexible redetermination provides for upward or downward price adjustment.

Fixed-price with incentive. This type of contract provides the seller with an incentive to control costs by establishing a target cost, a target profit, a ceiling price, and a final profit formula that allows a supplier to participate in any cost savings that accrue below target costs, thereby increasing supplier profits. Typically, if the seller can lower its costs, the buyer and seller share in the savings at prenegotiated rates, with the seller's profit increasing as costs go down. The greatest application for this contract form is in high-cost, long lead-time projects.

Firm fixed-price per unit level of effort. These contract forms are used when neither the work nor results can be specifically defined before performance. The parties therefore agree on a specific level of effort at a specific price (e.g., number of hours of testing at a fixed rate per hour), after which they will assess the results and decide if additional effort is required. Typical applications for such contracts include research and development work and laboratory testing.

Fixed Decreasing Prices

The learning curve or manufacturing progress function phenomenon occurs when, with increasing volumes, organizational learning should result in lower costs per unit. Used particularly effectively in the aircraft industry in World War II originally, it has also been used extensively in the electronics and automotive sectors. Under a fixed decreasing price contract a supplier would agree, for example, to a price of $1.00 per unit in year one of a contract, 97¢ per unit in year two, and 95¢ per unit in year three.

Strategic Joint Ventures, Strategic Alliances, Supplier Partners, and Preferred Suppliers

Long-term contracts provide for the development of joint ventures, strategic alliances, preferred suppliers, or supplier partnerships. Although these terms tend to be used interchangeably, they may show a gradation in terms of commitment by each side to the other. For example, in a joint venture or strategic alliance, significant investments may be made by one or both parties for research and development, market development, technology sharing, or other strategic initiatives. Supplier partnerships may require far less financial investment but substantial time commitment to relationship building and strengthening as well as commitment to continuous improvement. For some people, preferred suppliers are exceptional suppliers who are better than the norm and deserve special treatment.

Systems Contracting

An early type of long-term contract with which purchasing has had considerable success is systems contracting. In fact, systems contracting may be seen as the forerunner of alliances and partnering arrangements.

In the 1960s Ralph Bolton at Carborundum was an early proponent of systems contracting. The addition of modern technology with bar coding and EDI can make systems contracting particularly cost effective and responsive.

This type of ordering is most frequently associated with the procurement of maintenance, repair, and operating (MRO) materials. Its distinguishing feature is the high degree of integration between buyer-seller operations, which typically results in a net reduction of inventory for both. The contract itself often authorizes designated employees (generally from using departments) of the buying firm to place orders with the supplier for specified materials as needed while the contract is in force. This is accomplished by means of an order release form, which is frequently designed so that it can also be used by the supplier for stock picking and order assembly, as well as instructions for routing and delivery.

Such contracts can be written only after careful analysis and study by both buyer and seller of their respective operations. The seller must have materials available in the specified qualities and quantities when an order release is received from the buyer. To do this, however, the seller must have accurate and sufficiently detailed usage estimates from the buyer in advance of demand so that the seller can adjust production and its own procurement schedules to meet them. While this system is frequently referred to in the trade press as "stockless purchasing," and in principle has the capacity to be that, in practice, inventories are reduced (along with paperwork) but are seldom eliminated.

Just-in-Time Arrangements

Just-in-time is essentially an industrial engineering concept particularly focused on inventory reduction and value-added activities. In a typical "pull" system (further described in Book 3 in this series) a withdrawal from a subsequent operation or a withdrawal from stock "pulls" its replacement through the preceding operation. The triggering device may be a kanban, an empty space, or an order. To many purchasers JIT has meant transferring the responsibilities for inventories to suppliers. Even though there are substantial benefits to this—space, capital tie-up, the supplier stocking at its manufacturing costs rather than the selling price—this is not the optimal solution because the supplier's cost still needs to be reflected in the price paid. The idea of JIT coupled with continuous improvement is to remove the reasons for carrying inventory, such as reducing uncertainty, breakdowns, and quality problems so that inventory can be reduced throughout the value chain.

EDI

Electronic ordering systems with bar codes and electronic information and funds transfer permit fast, accurate, and low variable-cost solutions to many traditional supply challenges. EDI is discussed in detail in Book 4 in this series. Experts claim EDI's speed of adoption has been delayed by the fax machine, which permits fast transfer of standard forms without the pain of electronic conversion.

LETTERS OF INTENT

A letter of intent is a preliminary contract that may be used between buyer and seller to confirm certain agreements in connection with a procurement action. It serves as an interim purchase order or contract and provides immediate documentation of the salient features of an agreement. It anticipates that it will be superseded by subsequent documentation containing the same information with additional terms and conditions. The purpose of a letter of intent is to gain time in a commitment to a supplier prior to the issuance of a more complete purchase order or contract. Situations where a letter of intent may be used include the following:

1. To recognize that long lead times exist for the purchase of equipment, raw materials, design, training or hiring of people. One or both parties need to make commitments before contract completion or approval by a contracting board or a board of trustees or directors is possible.

2. To base volume discounts on future business without giving a firm order for the whole quantity. (Note that prices are usually

renegotiated if actual volume does not satisfy the discount structure.)

3.   To encourage a seller to stock items of interest to a buyer.

4.   To reserve a "place in line" for standard equipment when the formal purchase document approval system may require more time.

Types of Letters of Intent

There are two types of letters of intent: binding and nonbinding. The distinction depends primarily on the language used in the letter.

Binding Letters of Intent

A buyer may be bound by a letter of intent even if he or she doesn't wish to be unless steps are taken to prevent it. For example, if both parties sign a memorandum confirming negotiations or agreements in principle, the parties would likely be bound by contract unless the memorandum clearly and explicitly states that neither party intends to be bound and that this memo is not intended to be a contract. The same is true of minutes of negotiations meetings, as the parties can be held to have a contract for the points on which they have already agreed. Indeed, memos of agreement or negotiations that are expressly made conditional on legal approval may be viewed as a contract on the premise that legal points not approved by a company lawyer can still fall under the code for clarification.

Any letter of intent given as authorization to begin producing goods or the like will certainly be regarded as binding.

Nonbinding Letters of Intent

If a letter of intent is to be nonbinding, it must pass the test of explicit and clear statement to that effect, with the understanding that such a letter cannot be binding on one party and nonbinding on the other.

A letter of intent between the parties without authority to bind their companies would typically be held as nonbinding if it contains a condition of approval by management or other authority. However, one needs to be careful that the party truly does not have such authority and that there is no precedent of having treated prior, similar letters as binding, in which case the party may be considered possessive of "authority by ratification."

The best practice is to have the letter spell out in simple language whether or not it is intended to be binding.

Subcontracts

Subcontracts are legally treated exactly the same as contracts. Subcontracts are normally entered into by a prime contractor who "subs" out some or almost all of the work. For example, in construction, foundation work, electrical work, plumbing, painting, flooring, bricklaying, and roofing are only some of the subcontracts that can be let. In military, aerospace, and electronics industries, subcontracts are widely used. The complexity and technological sophistication of large government-funded projects will require contract administration officers whose task is to ensure that subcontracts are completed as agreed and that all documentation conforms to government requirements.

The Right to Subcontract
In government contracts sellers are normally obliged to give advance notice of the intent to subcontract and obtain consent before subcontracts are issued. Typically, the subcontractor clause is enforced in all cost-plus contracts and in fixed-prices with incentive contracts. If a primary contractor's purchasing system is government-approved, certain parts of the subcontractor clause may be waived.

In private business, buyers will often require the right to prior approval of subcontractors, but in the absence of such a requirement, suppliers are normally permitted to subcontract.

Notice of Awards

In private business, whether a notice of award is sent to all bidders typically depends upon the firm's policy and on the nature and value of the pending contract. Such procedures may be common practice for large dollar value or construction contracts but not for more routine contracts. The key is that the firm can decide to engage in total notice and debriefing based on what is considered good purchasing practice.

While the statutes are similarly silent for state and local public purchasing, most entities adopt standard practices for notice of awards to all bidders (successful or unsuccessful). However, in all public purchasing, except when a bidder gives justifiable advance notice that a part of a bid is proprietary, bid and award records are public. Therefore, a bidder has a right to review the records upon request.

Notification and Debriefing of Unsuccessful Bidders

Resolution of protest procedures vary for private, state, provincial, local, and federal procurement. In private organizations protest resolution is generally an administrative matter unless there is sufficient damage to a party to warrant resolution under an applicable arbitration clause or through litigation.

If not selected, suppliers are entitled to a reasonable explanation. However, in the private sector, care should be taken not to disclose the information contained in competitors' bids. In the public sector, bid information is normally accessible by the public. General information provided to the supplier will assist in better meeting the needs of the buyer during the next quotation process. As with all purchasing documents, files containing bids should be locked to prevent unauthorized access to the information.

Protests from unsuccessful bidders deserve quick and courteous treatment. Sometimes in larger public organizations, protests can be directed at a contract review board. In the public arena, unsuccessful bidders often try to enlist the assistance of their elected representatives to have a decision overturned. The resulting political pressure can be enormous. The procurement process used must be beyond reproach and in conformance with established policies and procedures. Adequate documentation has to provide proof that the award decision was appropriate. If, upon investigation, the appellant is found to have a sound case, rapid action is required to reverse the earlier decision, if this is still possible. Public contracting regulations have become more challenging over the years with potentially conflicting legislative requirements. Moreover, international trading requirements are trying to address restrictive public contracting rules.

In public organizations, statutes in some states and provinces require formal resolution under the Administrative Procedures Act, which calls for decision by an administrative judge with justifiable appeals to the courts. In other states or provinces, resolution decisions are made at the local and state or provincial level, as appropriate. In the United States, federal government contracts contain a dispute clause requiring that all disputes be resolved under the clause rather than in the courts. Typically, a decision is made by the contracting officer, with appeals directed to the Board of Contract Appeals or to the Court of Claims.

FOLLOW-UP AND EXPEDITING

The two terms *follow-up* and *expediting* are often used interchangeably, but in reality they refer to two quite separate activities after a purchase order has been placed or a contract agreed to. Follow-up is checking to ensure that the originally agreed-to schedule or delivery date will actually be honored. Expediting is an attempt by the purchaser to obtain earlier delivery than what was originally agreed to.

Reasons for Follow-Up and Expediting

Circumstances arise where follow-up and expediting may be required. It should be noted here that follow-up, in particular, adds no value and hence should not be necessary if the purchaser can be confident that the supplier will perform as promised. Presumably, it is not good practice to place orders with suppliers whose delivery performance is suspect. Total customer satisfaction and quality improvement programs have aided substantially in eliminating sloppy supplier attention to delivery promises made to customers. Reducing the need for follow-up not only helps reduce the purchaser's expense of staffing for follow-up but also helps reduce safety or buffer stocks carried because of delivery uncertainty.

At best, therefore, follow-up should be a matter of management by exception rather than a standard routine for all orders placed. In the case of open orders, a triggering device such as filing by a predetermined date can generate follow-up procedures. For critical or strategic contracts and with new or untried suppliers, follow-up is prudent to assure on-time delivery. In the section on major contract administration near the end of this chapter, it is easy to see how extensive such follow-up can be for complex major government or defense contracts.

Expediting

The need for expediting can arise because of unforeseen circumstances affecting the purchaser, the seller, or both. The more common situation affecting the purchaser is a change in plans requiring earlier delivery than originally required. In all fairness, if the reason for expediting originates with the purchaser and if the supplier has to incur significant charges to speed up the order, the extra costs should be borne by the buyer. Many purchasers prefer to avoid such legitimate charges and use the willingness of the supplier to deliver earlier as a test of the supplier's goodwill, flexibility, or responsiveness. Poor planning by the purchaser may result in constant expediting, a costly and unprofessional way to deal with suppliers. Just as a purchaser has a right to expect a supplier to deliver on time, a supplier has the right to expect a purchaser to commit to his or her delivery requirements.

Sometimes the purchaser needs a smaller or larger quantity than originally forecast. Delivery of a smaller quantity than originally forecast may not present substantial problems, but it may increase the cost per unit because of fixed setup costs, higher transportation charges, or the generation of surplus material. Delivery of larger quantities may not be done to original schedules or to original cost for a variety of obvious reasons. Short-cycled requests from buyers for engineering change orders, strikes, or other unforeseen circumstances need to be handled on an ad hoc basis.

Suppliers are normally well aware of the seriousness of late delivery and will try to do their utmost to correct the late situation on their own initiative. Often significant penalties exist to add further pressure on the late supplier. Under such circumstances, purchasers expect a supplier to sound the early warning as soon as possible to give the purchaser the maximum time possible to plan for the expected shortfall.

Just as it is reasonable for the purchaser to pay for extra charges when the purchaser wishes to deviate from plan, a supplier should recognize the extra costs a purchaser may incur as a result of late deliveries. The problem of consequential damages arises with extra charges that far exceed the value of the supplier's shipment or even the total business done with a purchaser. A late delivery of fasteners can shut down an automobile assembly line; a late delivery of concrete for the foundation can delay a whole building project. Even if the supplier can be negotiated into substantial penalty clauses for late delivery as part of the original contract, it is doubtful that these will cover the full cost of the delay. Therefore, the prime intent of substantial penalties is to ensure on-time delivery rather than financial recompense.

Expediting may also result from a supplier's circumstances. The original delivery promise may turn out to be unrealistic or too difficult to meet; supply difficulties or equipment breakdowns may prevent on-time delivery; or goods may be damaged or stolen in transit. If a supplier has back orders, expediting can be used to ensure earliest possible delivery of that portion of the original order not delivered on time. When the total order is late, orders that are critical may require constant attention, while those not yet critical may require a lower intensity of expediting. In a just-in-time environment, late deliveries can have very serious consequences for the purchaser.

RECEIPT

It is important to receive incoming shipments properly. In most organizations, receiving is centralized and reports to purchasing. Incoming shipments need to be checked to ensure that (1) the goods were ordered, (2) no visible damage occurred in transit, (3) the quantities received are correct, and (4) the shipping documentation is correct and complete.

A receiving report is normally used to record the receipt, confirm the delivery, help close the order cycle, and ensure payment to the supplier.

Under JIT arrangements in assembly industries, it is not unusual to have shipments delivered directly to the point of use in the plant, bypassing traditional receiving processes and recordkeeping. Book 3 in this series covers receiving more extensively.

INSPECTION

After a purchase commitment has been made, inspection may be required to ensure that what is delivered conforms to the original description or specification. Since inspection does not add value, the desirable situation is one in which no inspection at all is necessary. The supplier and purchaser may have cooperatively mounted quality assurance programs resulting in outstanding supplier quality performance and records. Thus, the purchaser recognizes that inspection at the purchaser's premises would provide no additional quality information useful to the purchaser. Such quality confidence is vital to the effective performance of a just-in-time system, where supply interruptions because of quality problems would be intolerable.

Despite numerous quality programs, the Baldrige Award, the ISO 9000 series of standards, and the efforts of quality gurus like Deming, Juran, and Crosby, there is still a need for inspection to assure quality conformance. It is normal to watch new suppliers or untried or new products with special care. Moreover, spot checks of established suppliers assure continuing quality vigilance. There is no point in spending time and money on preparing satisfactory specifications unless adequate provision is made to ensure that specifications are lived up to by suppliers.

Under the UCC, no provisions exist for just-in-time arrangements or deliveries from certified suppliers where no incoming inspection is done. Periodic quality audits fall into the same class. Therefore, the contract should document such quality assurance arrangements between buyer and seller with both parties as signatories.

Specifications should include the procedure for inspection and testing as protection for both buyer and seller. This avoids potential disagreements. While in some situations purchasers may be more sophisticated in quality assurance and in others suppliers may be, it is sensible for both sides to cooperate on this issue.

Right to Inspect, Return, and Withhold Payment

The buyer has the right to inspect goods before acceptance, provided that this is done within a reasonable period. When a purchaser accepts merchandise after inspection, he or she is ordinarily prevented from subsequently raising an issue with respect to quantity or quality. Acceptance is the buyer's assent to take ownership of the goods provided by the seller. Any words or actions that indicate buyer's acceptance are sufficient.

If a buyer keeps the goods and acts as if he or she accepts them, then acceptance has taken place, even if the buyer makes statements to the contrary.

If the goods fail in any way to conform to the contract, the buyer has the option to (1) reject the whole shipment, (2) accept the whole shipment, or (3) accept only part of the shipment. If the buyer rejects the goods, he or she must hold them with reasonable care until the seller has sufficient time to remove them. The buyer can take the following actions:

Withhold payment. When a contract calls for payment before inspection, payment must be made unless the delivered materials are so obviously nonconforming that inspection is not needed. But payment required before inspection does not constitute final acceptance of goods. Rejection can still occur if the buyer inspects after the required payment and finds the goods to be unsatisfactory.

When a seller delivers nonconforming goods, the buyer can also recover any prepayments made to the seller.

Dispose of nonconforming goods. If, upon inspection, goods are found to be nonconforming, and the time allowed for seller performance has passed, the goods may be disposed of in one of the following ways:

Accept the goods. They may be accepted as they are or with reasonable expectations that the seller will cure the nonconformity.

Reject the goods. The goods may be rejected outright, upon which they are returned to the supplier (at the supplier's expense) on a shipping order and invoice issued by the purchasing department. The supplier is notified of this action and the reasons thereof. It is also determined at this point whether the original purchase order is still in effect or whether the order is terminated through the default of the supplier. An alternative procedure is to return the material for replacement, which is frequently done in the case of fabricated parts.

Request rework. Frequently, suppliers will have representatives come to the buyer's plant to make necessary adjustments to faulty equipment or work out a satisfactory application of the nonconforming materials.

A buyer is obligated by law to notify the seller of his or her intention to either accept nonconforming goods contingent upon seller cure, reject nonconforming goods, or request rework of nonconforming goods. If the seller is notified of the buyer's rejection of goods or intention to rework goods and does not respond to it within a reasonable amount of time, the buyer may:

- Store the goods and claim for storage charges.
- Reship the goods freight collect.
- Resell the goods and deduct reasonable sales costs from the proceeds.
- After the seller has inspected the goods or waived the right to inspect, deduct the cost of reworking the goods from the price of the goods.

Revocation of acceptance. A buyer may revoke acceptance of goods when they fail to conform to a contract to such an extent that the defects substantially impair their value, provided that he or she accepted the goods without knowledge of their nonconformity or that he or she accepted the goods with knowledge of nonconformity but had reason to believe that the seller would cure the default. In other words, a buyer cannot revoke acceptance merely because the goods do not conform to the contract, unless the nonconformity substantially impairs their value to the buyer. Revocation of acceptance may be made with respect to the entire quantity of goods or to a particular portion. A buyer who revokes acceptance stands in the same position as though he or she had rejected the goods when they had been originally tendered. A buyer must give a seller notice of revocation, and a seller has a reasonable amount of time to attempt to correct the defects in the goods.

Invoice Checking and Discrepancies

It has always been considered good practice to match the invoice received to the purchase order, receiving report, and waybill to ensure that no discrepancies exist. Purchasing has traditionally claimed it should do the checking since it issued the order. Accounting claimed it should check because it looked after payment. Regardless of where the checking is done, discrepancies need to be resolved quickly.

Invoices that do not coincide with the shipment or the purchase order require the buyer to determine if it is the fault of the carrier or the supplier. In either event, the responsible party should be notified as soon as possible in order to get the necessary details verified.

It is the responsibility of the buyer to contact the parties so that the details can be ironed out and the supplier and carrier can be properly compensated. Discrepancies can take the form of any one of the following: (1) late or early delivery; (2) incorrect quantity or part shipment; (3) unacceptable quality; (4) goods damaged in transit; (5) incorrect pricing; and (6) other errors or problems. Prolonged payment problems may result in a "credit hold" being put on the purchaser's account. Normally, the supplier will not ship additional product or provide services while the credit hold is in place. The hold is lifted when the proper compensation is made.

It is common practice to debit the account of a supplier for a product found unacceptable during the warranty period. After the supplier takes proper corrective action (repair or replacement of the goods), a credit memo is forwarded to offset the debit.

Risk of Loss and Delivery and Transportation Issues

A large number of issues can potentially arise in the transportation of materials, goods, or equipment from the supplier to the purchaser. Since the third textbook in this series deals extensively with transportation concerns, only several will be raised at this point.

Risk of loss disputes, as they relate to a transfer of goods between two parties, generally arise in two situations. The first is a direct transaction between two parties. In this situation, the party in possession of goods has the best opportunity to protect them and therefore bears the risk. Here risk of loss passes with delivery, unless the seller is in breach of contract. In such a case, the buyer has a right to rejection, and the seller retains risk until the nonconformity is cured.

The second situation involves a third party (i.e., carrier). Here risk of loss depends on the FOB (free on board) rules:

FOB origin. The seller bears risk until it loads the goods onto an appropriate carrier, after which the buyer assumes risk of loss and must claim against the carrier for damage or loss in transit.

FOB destination. The seller bears risk until the goods are transported to the buyer's dock, after which risk will pass to the buyer.

FAS. This term, used when goods are transported by ship, stands for free along side. It requires that a seller place the goods on a loading dock accessible to the carrier, at which time risk of loss transfers to the buyer.

CIF. This international shipping term stands for cost, insurance, and freight, and is typically used along with a destination point. For example, a shipment CIF buyer's plant would obligate the seller to deliver to and load the goods onto a carrier, obtain a bill of lading, pay freight, insure the goods, prepare necessary documents, and deliver to the buyer all documents necessary to receive the goods.

Liability associated with in-transit goods may extend far beyond the traditional risk of loss as determined under FOB rules. For example, an in-transit spill of hazardous material shipped FOB Origin could be very serious for the buyer's firm, because liability will generally follow title, and title would have transferred to the buyer at the origin. Common sense tells us that the buyer's firm is better insulated from liability under FOB destination contracts on all such goods, and that carriers should be adequately insured for the specific hazards involved.

Buyers should remove any ambiguity in FOB points by clearly designating exactly where the risk of loss is to transfer. Suppose that a contract for delivery at the buyer's plant in New York City was written

FOB New York and that loss occurred just inside the city limits. A case could be made that the destination point was reached and that the buyer had risk of loss, though the goods never actually reached him or her. Similarly, if a seller has goods shipped from some other point, FOB origin contracts may involve shipping costs much higher than those from the address to which a contract is made.

Concerns relating to delivery and risk of loss include:

- *Seller cure for in-transit loss.* One might reason that sellers will readily replace goods lost in transit where they have the risk of loss. However, in case of such problems as shortages, it may be in the seller's best interest not to replace them rapidly. Accordingly, it is good practice to include in the contract how long a seller can take to cure for in-transit loss of goods.
- *Piecemeal deliveries.* An equally important item to include in a contract is that delivery is not considered complete until the buyer receives everything needed to test or use the goods or equipment.
- *Use of buyer-specified carrier.* It is common practice in today's environment for buyers and their traffic departments to negotiate freight rates on measures like volume. These can entail considerable discounts. Accordingly, sound practice dictates buyer specification of the carrier and a systematic procedure of warning and charge-back when freight charges are higher due to seller failure to follow instructions.

PAYMENT

Purchasing, in its simplest form, is the exchange of money for goods and services. The seller is responsible for providing goods and services as agreed in the contract. The purchaser is responsible for payment as agreed in the contract.

Today's exacting purchasers are placing additional demands on suppliers that were not common over a decade ago. TQM, continuous value improvement, JIT, ESI, EDI, supplier base reduction, certification, systems contracting, and integrated logistics, to name just a few, require a high degree of conformance and precision of the supplier. It is not at all unusual for a purchaser to demand that a lot of exactly 1,000 perfect quality units, not 999 and not 1,001, be delivered to one certain place between 2 and 3 pm in the afternoon and that the lot be packaged in a specific way. On top of this, significant penalties are involved for any discrepancies. It only makes good business sense that the purchaser should ensure that payment of the supplier be managed with equal accuracy and precision. The excuse that accounts payable is responsible for payment is a hollow one from the supplier's perspective.

The absolute first requirement in a solid buyer-supplier relationship is the assurance of supplier payment as promised in the purchase agreement. This does not always have to mean prompt payment. Quick payment, quicker than trade practice, should deserve an early payment discount. The issue here is one of contractual obligation. It does not matter whether payment needs to be made in 10, 30, 60, 90, or 120 days. Whatever terms and conditions have been agreed to need to be honored by the purchaser. Failure to do so prevents the establishment of a sound buyer-seller relationship.

Normally the payment cycle commences upon delivery of an acceptable product or successful completion of a service. Errors in receiving, invoicing, or purchase order processing delay the payment. Both the purchaser and supplier should make a good faith effort to resolve discrepancies in a timely manner.

The purchase contract should specify at what time the payment clock starts ticking. Cash on delivery (COD) is fast. Payment terms may start on the day of shipping from the supplier's operation, day of receipt at the buyer's premises, or at some other mutually agreed point. For example, for equipment that is provided under a performance specification, a certain portion or all of the payment may be triggered by the date on which the equipment, once installed and operating, can reach a predetermined output of satisfactory quality. Standard payment terms of net 30, 60, or 90 days recognize that it is to the buyer's advantage to be able to use the supplier's cash to help finance the buyer's operations. A retail store, for example, which has 60 days from receipt of goods to pay, but which can sell these goods within 10 days, has the use of the supplier's cash for 50 days.

In some organizations the financial strategy is to balance the accounts receivable against the accounts payable. Clearly, payment terms should be part of the total value of the deal and can be construed as a form of price discount.

Early payment incentives such as 1 or 2 percent are offered by suppliers to encourage quick payment of invoices. The taking of these discounts without observing the time constraint, such as taking a 1 percent discount but still paying in 30 days, is unethical.

Checking invoices and preparing checks is in many organizations expensive, in some cases a cost of as high as $150 per check issued. Therefore, many organizations attempt to minimize the number of invoices submitted for payment and checks that need to be issued. Such options include reducing the supplier base; the use of systems contracts; blanket orders; weekly, bimonthly, or monthly billings; invoice consolidation; the Kaiser check system; credit cards; and electronic funds transfer.

Automatic payment without invoices eliminates the need for activities that add no value and is particularly appropriate for JIT applications. For example, an automobile assembly plant that produces 1,000 cars per day uses 1,000 steering wheels per day. Since the exact number of vehicles assembled is well known there is no need for the supplier of steering wheels to submit an invoice for the 1,000 steering wheels (and in a certified quality environment, no need for receiving report, inspection report, moving slip, waybill, returns, and any other documentation that might traditionally have accompanied such a delivery).

Attempts to apply the value chain concept, life cycle costing, and reengineering to the procurement cycle identify the many costs incurred inside and outside the procurement area as a result of orders or contracts issued. Every purchaser should be aware of these extra costs—in this case as related to payment, but in others it might be receiving, quality assurance, inventory, or operations.

Progress or Milestone Payments

An effective method for providing suppliers with interim compensation during a long-term contract is to offer progress or milestone payments. To minimize the risk associated with offering such payments, the purchaser should tie them to tangible events that are easily verifiable. Mutually acceptable criteria should be established in writing at the onset of the contract in order to avoid any confusion as to what events trigger payment to the supplier. The contract may stipulate that satisfactory completion of the entire task is required or else the customer may recoup funds already expended.

Not linking payments to discernible events presents a significant risk to the customer. The supplier may be receiving payments but may not be proceeding in a manner that executes the contract on a timely basis.

Cost Price Overruns

In a fixed-price contract, the purchaser is not obligated to compensate the supplier for cost overruns. However, supplier performance may degrade as the loss escalates. The purchaser should consider the following parameters when presented with a request from the supplier for compensation for cost overruns:

- The financial viability of the supplier.
- The possibility of getting the job completed through another source, and the associated cost of doing so.
- The investment made by the customer in the project, and whether or not it is recoverable from the supplier.

RECORD MAINTENANCE

After having gone through the steps described, all that remains for the disposal of any order is to complete the records of the purchasing department. This operation involves little more than assembling and filing the purchasing department's copies of the documents relating to the order and transferring to appropriate records the information the department may wish to keep. The former is largely a routine matter. The latter involves judgment as to what records are to be kept and for how long.

Most companies differentiate the importance of the various forms and records. For example, a purchase order constitutes evidence of a contract with an outside party and as such may well be retained much longer than the requisition, which is an internal memorandum.

The minimal basic records to be maintained, either manually or by computer, are:

1. A PO log, which identifies all POs by number and indicates the open or closed status of each.

2. A PO file, containing a copy of all POs, filed numerically.

3. A commodity file, showing all purchases of each major commodity or item (date, supplier, quantity, price, PO number).

4. A supplier history file, showing all purchases placed with major suppliers.

Some of the optional record files include:

1. Labor contracts, giving status of the union contracts (expiration dates) of all major suppliers.

2. Tool and die records showing tooling purchased, initial life (or production quantity), usage history, price, ownership, and location. This information may prevent the buyer's being billed more than once for the same tooling.

3. Minority and small business purchases, showing dollar purchases from such suppliers.

4. Bid award history file, showing suppliers asked to bid, amounts bid, number of bids, and the successful bidder, by major items. This information may highlight supplier bid patterns and possible collusion.

Record maintenance is covered more extensively in Book 2 of this series.

DISPOSAL

Disposal has become an important issue in the acquisition process. Environmental legislation and the needs for cost reduction and economic recovery have together put disposal in the limelight. In many organizations the purchasing area is responsible for disposal. Categories of materials for disposal include excess or surplus materials, obsolete material or equipment, rejected end products, scrap material, waste, and hazardous waste. Books 3 and 4 in this series cover disposal and also identify ways to maximize the savings potential.

It is difficult to avoid disposal issues in purchasing, even if it is just for the packaging for incoming requirements. Therefore, disposal considerations become important at the time of need definition and should definitely affect the evaluation of the best acquisition option and the supplier chosen.

MAJOR CONTRACT ADMINISTRATION

In the broadest context, contract administration refers to all activities pertaining to a particular contract by buyer and seller from contract award to contract closeout. The purpose of contract administration is to ensure that (1) the buying organization lives up to its contractual commitments and (2) the selling organization fulfills its obligations under the contract. This is partially accomplished by prompt and fair resolution of any problems that arise, negotiating equitable adjustments when warranted, and ensuring proper documentation of all contractual transactions.

Since governmental contracts are often more exacting than private contracts, particularly in terms of documentation and contract changes, they form a good model for sound administration practice. Typical steps in routine major contract administration include (1) work control; (2) assuring compliance with contract terms and conditions; (3) assuring financial control; (4) monitoring, auditing, and approving contractor systems; and (5) other administrative actions.

Work Control

Some types of contracts (such as indefinite delivery contracts, time and material/labor hour contracts, and cost reimbursement contracts) defer the ordering and work authorization processes until after the award. In these circumstances, ordering becomes a postaward or contract administration matter.

The customer activity initiates a standard work order with the line item description of the desired items copied directly from the contract, along with the unit price. The desired quantity is multiplied by the unit price to arrive at a total. This work order is generally routed to either a buyer representative or buyer for signature and distribution to the supplier, the customer, and the finance office.

Work completion, work inspection and acceptance, invoicing, and payment follow normal procedures. For time and material/labor hour contracts the customer-prepared statement of work is generally accompanied by an in-house estimate.

Cost Reimbursable Contract Work Control
Cost reimbursement contract work control procedures are often considerably more complex. Many cost reimbursement contracts require the establishment and maintenance of a process of annual work plans, work authorizations, and notices to proceed (NTPs) as a means to assist in cost and schedule control.

Upon buyer receipt of preliminary funding guidance, the supplier can be provided the expected levels of funding for the ensuing (budget) and subsequent fiscal years, milestones based on the current master program schedule, and relevant scope information. This guidance enables the supplier to prepare an annual work plan (AWP).

The AWP is central to the total process; it provides the initial definition of tasks to be performed in the budget year and a schedule for accomplishment. The AWP balances funding guidance and program schedule requirements. During the AWP review, the supplier resource projections are approved and the tasks to be undertaken are scheduled.

The specific elements of an AWP generally include goals and assumptions; work authorization review results; and a schedule, staffing plan, and cost estimate for the budget year.

Work authorizations generally contain a scope of work, including work breakdown structure designations for the work; information regarding the duration of the work authorization; the baseline cost estimate for the work; and references to the existing AWP and notices to proceed (NTP) to be issued subsequent to the work authorizations.

Tasks to be accomplished during a certain period or phase of contract performance will be described in NTP documentation and issued to the supplier prior to the supplier's undertaking any work. The NTP normally includes a statement of work, the key schedule milestones for task accomplishment, and the total amount of funds allotted to the tasks. Upon receiving the NTP, the supplier will begin work and start cost and scheduling reporting for the tasks concerned.

Assuring Compliance

A contract is a tool that the buyer uses to acquire goods and services; contract administration is an attempt to manage this tool effectively so as to obtain what is needed. Sometimes corrective action will need to be taken as part of effective contract management. Since such action cannot be initiated unless the nature and extent of problems are known, the buyer must have mechanisms for identifying problems and determining their significance to the contract objectives. Those mechanisms constitute contract compliance monitoring or surveillance. Contract compliance monitoring is the responsibility of the entire contract administration team; it aims to determine whether:

- Performance on schedule can be expected.
- The cost will be within the estimate.
- Resources are being applied at originally predicted levels.
- The quality of the end products will be consistent with the specifications.
- Progress payments are warranted.
- New components need to be incorporated in major equipment.
- The contractor's own progress monitoring system is adequate.
- All contractual provisions (including those not relating to the work itself) are being followed.

Assuring Financial Control

From the award of a contract to final closeout the supplier's primary concern is to receive payment for work done and for funds expended in as timely a manner as possible. The different types of contracts used by the buyer create different financial relationships between the buyer and the supplier. The supplier with a firm fixed-price contract has a strong incentive to perform in the most economical way, since every penny saved below the contract price is additional profit. Under most labor hour, time and materials, and cost reimbursement contracts, however, the supplier has little incentive to perform in the most economical way.

Under these latter types of contracts, the supplier is generally entitled to compensation for either a fixed amount per hour or for the costs incurred in doing the work, provided that the expenses are not unreasonable. Moreover, the work description in such contracts is usually broad because it is difficult to predict just what the contractor is required to do. This gives the supplier broad contractual authorization, which can permit the supplier to perform and charge for effort along lines other than those specifically wanted.

Therefore, the buyer needs to monitor and guide the supplier's efforts in order to prevent waste of funds and to ensure that the purchaser gets the services needed within the amount budgeted.

Monitoring, Auditing, and Approving Contractor Systems

The federal government relies upon a number of formal programs to determine whether its major suppliers conform to law, regulation, good business practice, and the "federal norm." Many other governmental, quasi-governmental, and institutional clients impose similar, although generally less rigid, requirements. This section describes some of the key requirements imposed on federal government prime suppliers, particularly those that contract with the Department of Defense (DOD).

The most common of these formalized programs is the subcontract consent review process, explained in the subcontracts clause of most major contracts. This clause requires suppliers performing other than firm fixed-price contracts to follow the federal norm in their award of subcontracts and purchase orders, and to request specific written consent to place large subcontracts and purchase orders. Close scrutiny of subcontracts and purchase orders is generally necessary in the following circumstances:

- When the supplier's purchasing system is considered inadequate.
- When subcontracts are awarded without competition or the proposed prices seem unreasonable and the dollar amounts involved are large.
- When close working arrangements or affiliations with subcontractors may result in higher than normal subcontract prices.
- When the price proposed by the subcontractor is less favorable than prices quoted by the subcontractor for comparable jobs.
- When the subcontract is placed on other than a firm fixed-price basis. Buyers should be especially skeptical of the repeated award of subcontracts on a cost reimbursement or time and materials/labor hour basis.

A typical subcontracts clause permits the buyer to perform contractor purchasing systems reviews (CPSRs) whenever a prime supplier's negotiated defense sales are expected to exceed $10 million. This type of review samples the subcontracts and purchase orders awarded by the prime supplier in order to draw conclusions about the degree to which the supplier has protected the taxpayers' interests.

Other Administrative Actions

Other contract administration issues that need to be addressed include (1) approving costs or adjusting prices; (2) negotiating change orders; and (3) supplier relationships and communications issues.

Price Adjustment

If the contract contains price adjustment clauses, the contract administrator must ensure that contractual terms are executed properly. Typical price adjustment clauses may fall into one of the following three categories:

1. *Established price clauses.* These clauses appear in different forms depending on the types of materials purchased and the types of suppliers involved. In all cases, however, adjustments are made in

two ways: (1) in accordance with the fluctuations in the supplier's applicable established prices, or (2) in accordance with applicable labor and material price indexes.

2.   *Adjustment clauses based on actual cost methods.* This type of clause generally permits adjustments to labor or material costs where no major element of design engineering or developmental work is involved in producing the item being procured and where identifiable labor or material cost factors are subject to change. The contract schedule should describe in detail the types of labor or material subject to the actual cost adjustment and the costing method to be used.

3.   *Adjustment clauses based on published indexes.* The third type includes clauses based on adjustments to labor or material costs using a published cost index method. For several reasons, including ease of administration and the ability to base the adjustment on readily available published economic data, this type of clause is preferred over an actual cost clause. This type of clause should be used when:

   a. There will be an extended period of performance with significant costs to be incurred after the first year of contract performance.
   b. The contract amount subject to adjustment is substantial.
   c. The economic variables for labor and material are too unstable to reflect a reasonable division of risk between the parties without economic price adjustment provisions.

Change Orders
Ideally, a contract contains all provisions necessary for completion of the work and discharge of both parties' obligations. Modifications of the agreement are not contemplated when the contract is signed. In practice, however, few contracts are completed without some type of modification. Some modifications are simply administrative changes that do not affect the substance of the contract. Others involve substantial changes to the

price, quantity, quality, delivery, or other terms originally agreed to by the buyer and the supplier.

Obviously, if the changes are substantive, agreement must be obtained from the relevant internal or external customers as well as others who may be significantly affected. For example, a major price increase may affect a service, a product, project viability, or even the purchaser's ability to pay. On a happier note, an unexpected, timely technological breakthrough may offer an opportunity to lower costs, increase quantities, or change specifications.

Because the buyer's authority under any given contract is defined by the contract's clauses, those clauses should contain provisions that allow alteration of the contract after award. There should also be provisions that require the parties to alter the delivery schedule equitably, or the price to be paid in correspondence with other changes in the contract's terms. Consequently, contract provisions must give the buyer authority to make changes. These provisions should also allow the supplier or the buyer relief (if the other party does something not contemplated under the original agreement or fails to do something contemplated by the original agreement) as well as equitable adjustment of performance time or price when changes are made.

Supplier Relationships and Communications
Professional purchasers recognize that effective supplier relations and communications are vital to successful contract completion. In the governmental sector, contract and subcontract administration requires conferencing and progress reporting. In the private sector, strategic and longer-term contracts, service contracts, just-in-time contracts, EDI, single sources, preferred suppliers, and strategic alliances all require more attention to relationships and communications than the traditional short-term, competitively bid, standard purchasing orders. Recognizing the value of human input, information exchanges, mutual trust, and support allows for synergies difficult to achieve in other ways. Some relatively simple means include supplier conferences and progress reports. More complex interrelationships are covered in the fourth textbook in this series.

The foremost prerequisite to successful contract performance is a sound understanding by both parties of the contract requirements.

Supplier Conferences and Progress Reports

When the dollar magnitude, complexity, or importance of the work demands that buyers hold a conference with the prospective supplier either immediately prior to or immediately after the award of the contract, the buying organization's contract administration team will meet with the supplier's team. The issues to be discussed are addressed in technical terms in the request for proposal and the resulting contract. This conference is the vehicle the team uses to ensure that these provisions are fully understood and implemented.

Progress Reports

In some instances, the supplier is required by the terms of the contract to submit a phased production schedule for review and approval. A phased production schedule for a new custom manufactured product would show the time required to perform the production cycle—planning, designing, purchasing, tooling, plant rearrangements, component manufacture, subassembly, final assembly, testing, and shipping.

In most cases, the buyer may include a requirement for production progress information in the request for proposal and in the resulting contract. The ensuing reports frequently show the supplier's actual and forecast deliveries (as compared with the contract schedule); delay factors (if any); and the status of incomplete preproduction work such as design and engineering, tooling, and construction of prototypes. The reports should also contain narrative sections in which the supplier explains any difficulties and action proposed or taken to overcome the difficulties.

Production progress reports do not alleviate the requirement to conduct visits to the supplier's plant or the worksite on crucial contracts. The right to conduct such visits must be established in the RFP and the resulting contract. On critical contracts, where the cost is justified, it may be desirable to establish a resident plant monitor at the supplier's facility to monitor the quality and timeliness of the work being performed.

Supplier Feedback

An often overlooked aspect of successful supplier communication is the solicitation of supplier feedback. This process involves solicitation of supplier answers to the following types of questions:

- How knowledgeable are our buyers?
- How accurate are our engineering specifications?
- How clearly do we state our quality requirements?
- How timely are our payments?
- How receptive are we to suggestions?
- How timely are our responses?

Organizations that use supplier surveys usually find that the feedback plays a significant role in improving supplier relations and in ensuring that high-quality materials are received on time.

ETHICS

All business and government conduct requires adherence to legal and ethical standards. The procurement field is no exception, and over the years an impressive body of knowledge has developed to define the issues.

In the next chapter, the legal requirements as they affect contracts and remedies are highlighted. Many laws directly or indirectly affect the supply field. Aside from contract law, environmental legislation, social legislation (including equity and minority requirements), transportation, and international agreements such as GATT are just a few of the threads in the legal web hovering over buyers and sellers.

Ethics also provide rules of conduct but tend to go beyond requiring conformance to the laws of the land. Every profession has its ethical code representing values and beliefs about proper conduct that attempt to ensure that the members of the profession will behave properly with each other as well as with their employers, customers, and the general public. In professional fields such as medicine and engineering, public safety may be at stake. In professions such as accounting and purchasing, public protection from fraud is more at issue. In all professions, failure to observe the ethical code of the profession will affect the public trust.

Codes of ethics represent years of experience in the field, and adherence to them is not just a matter of enlightened self-interest. It is an obligation of everyone who aspires to be a professional.

The procurement field is rife with ethical issues. Because large sums of money are expended, large numbers of people are involved in the process on both the buyer and the seller side, and organizational competitive and environmental pressures may be extremely stressful, it is no surprise that ethical issues abound. In some parts of the world, bribery is still practised and is considered an accepted way of ensuring that goods and services will be purchased and provided. In most countries bribery is illegal, as are libel, slander, and disparagement. NAPM, the Purchasing Management Association of Canada (PMAC), and the National Institute of Governmental Purchasing (NIGP) provide a code of ethics for their membership and anyone else who wishes to practice in the field.

The following is a synopsis of the NAPM *Principles & Standards of Purchasing Practice*, January 1992.

Perception

Avoid the intent and appearance of unethical or compromising practice in relationships, actions, and communications.

- Consider a reassignment of buying responsibility if buying from a personal friend.

- Choose appropriate business meeting locations, and avoid noticeable displays of affection and excessively personal conversations.

Responsibilities to the Employer

Demonstrate loyalty to the employer by diligently following the lawful instructions of the employer, using reasonable care and only authority granted.

- Maintain up-to-date knowledge of laws, understand the authority granted, and apply the legal and ethical concepts of the agency relationship.

- Obtain the maximum benefits for monies expended, and promote competition, the analysis of least total cost, and the long-term best interests of the employer.

Conflict of Interest

Refrain from any private business or professional activity that would create a conflict between personal interests and the interests of the employer.

- Avoid using a company position to further outside activities, engaging in business, or business with a supplier that encroaches upon or competes with the employer.
- Report ownership of a supplier's stock.
- Use conflict of interest statements and self-evaluation procedures.

Gratuities

Refrain from soliciting or accepting money, loans, credits, or prejudicial discounts, and the acceptance of gifts, entertainment, favors, or services from present or potential suppliers that might influence or appear to influence purchasing decisions.

- Evaluate any gift or entertainment on its legality, the best interest of the employer, the influence it may exert on a buying decision, and the perception of others.
- Keep business meals to a specific business purpose, meet infrequently with the same supplier, and pay as frequently as the supplier does.

Confidential Information

Handle confidential or proprietary information belonging to employers or suppliers with due care and proper consideration of ethical and legal ramifications and governmental regulations.

- Develop a formal policy, divulge information on a need-to-know basis, label confidential information appropriately, and use formal confidentiality agreements.

Treatment of Suppliers

Promote positive supplier relationships through courtesy and impartiality in all phases of the purchasing cycle.

- Establish parameters for bidding, rebidding, and/or negotiating.
- Maintain confidentiality of a supplier's prices, terms, or proprietary information.
- Avoid unreasonable demands and make a prompt and fair resolution of problems.
- Return telephone messages promptly and courteously, and maintain a friendly, cooperative, and yet objective relationship.

Reciprocity

Refrain from reciprocal agreements that restrain competition.

- Oppose any form of corporate reciprocity as part of purchasing strategy.
- Be knowledgeable of antitrust laws, but seek legal counsel.
- Do not provide lists of suppliers to sales or marketing.

Federal and State Laws

Know and obey the letter and spirit of laws governing the purchasing function, and remain alert to the legal ramifications of purchasing decisions.

- Let legal counsel interpret the law, and provide preventive analysis and planning.
- Do not intentionally pursue loopholes in written laws.
- Understand and apply applicable government regulations.

Small, Disadvantaged, and Minority-Owned Businesses

Encourage all segments of society to participate by demonstrating support for small, disadvantaged, and minority-owned business.

- Adhere to applicable laws and regulations, and strive to attain policies and goals.

- Participate in organizations that stimulate the growth of these businesses, and encourage employees to seek new sources of supplies and services.

Personal Purchases for Employees

Discourage purchasing's involvement in employer-sponsored programs of personal purchases that are not business-related.

- If personal purchase programs exist, avoid buying for specific individual nonbusiness use; ensure that arrangements are fair to suppliers, employees, and employers; do not force special concessions on suppliers; and inform suppliers that such purchases are for employees.

Responsibilities to the Profession

Enhance the proficiency and stature of the purchasing profession by acquiring and maintaining current technical knowledge and the highest standards of ethical behavior.

- Support the development, recognition, reassessment, and application of technical and ethical standards.

- Achieve and maintain technical knowledge.

- Maintain the relevance of policies, practices, programs, and activities to the best interest of the profession.

International Purchasing

Conduct international purchasing in accordance with the laws, customs, and practices of foreign countries, consistent with US laws, your oganization policies, and these ethical guidelines.

- Be especially sensitive to local laws, customs, and cultural differences.

- Use your company chain of command, company counsel, the Foreign Corrupt Practices Act for guidance; recognize that US laws won't apply in most countries.

Normally each organization develops its own internal set of ethical guidelines for its employees. These often bear a significant similarity to the professional codes issued by the associations and may, in fact, include them. The key policies of Intel Corporation follow:[1]

Intellectual Property Right Protection

Intel respects the intellectual property rights of others—business associates, suppliers, and competitors alike. Intel is committed to the protection of a supplier's confidential information and insists on the same dedication to protection of Intel's disclosed confidential information. Furthermore, it is a requirement that Intel's mutually protective Corporate Nondisclosure Agreement or other nondisclosure agreement be signed by the supplier and Intel before the disclosure or exchange of any confidential information takes place. When Intel property rights are protected by patent, trademark, copyright, maskwork, or trade secret laws, Intel, by contract, provides infringement indemnification to its customers. Conversely, when Intel is the customer, it requires supplier infringement indemnification against intellectual property rights claimed by third parties.

Ethics

Purchasing will conduct itself ethically and fairly in relation to the suppliers with whom Intel does business.

---

[1]Courtesy of Intel Corporation.

Gratuities

Intel personnel and members of their families worldwide are prohibited from accepting gifts or gratuities of any form from current or potential suppliers. Acceptance of such offerings may be interpreted as an attempt to improperly influence decisions. Unsolicited advertising or promotional items with company logos are acceptable as long as their value is less than $25.

Reciprocity

While Intel may sell its products to its suppliers, it is against Intel policy to require a supplier to purchase our products as a condition for Intel doing business with that supplier.

Publicity

Intel maintains a strict policy that neither party may use the other's name in advertisements, press releases, design win announcements, or business level disclosures without prior written consent.

Small and Minority Suppliers

Intel appreciates the valuable contributions made by small, disadvantaged, and women-owned businesses. Therefore, a good faith effort is made to allow qualified small and minority businesses the maximum practicable opportunity to receive a fair share of Intel's business.

Controlled Substances

To ensure a drug-free and risk-free environment, suppliers are required to support Intel's policy that prohibits the use, sale, or possession of illegal drugs and alcoholic beverages on Intel premises or work areas.

Environmental, Health, and Safety

Intel expects suppliers to understand and fully comply with all applicable federal, state, and local laws, including, but not limited to, all environmental, health, and safety (EHS) and related laws. In addition, suppliers must agree to abide by all Intel rules, including but not limited to all applicable EHS policies, procedures, and guidelines. Further, to the extent that supplier obtains, uses, handles, or transports any hazardous materials (as defined by any law, standard, or practice) on Intel's property, or to perform work for Intel, supplier must warrant that they understand the nature and hazard associated with the use of any such materials. Supplier must agree to be fully responsible for any liability resulting from the supplier's use of hazardous materials.

Part of the difficulty of good ethics lies not in the development of suitable guidelines but in interpretation and adherence. Most purchasers would not dream of accepting a sum of money to induce them to do business with a particular supplier. That would be both illegal and unethical; yet the acceptance of tickets to sporting events, gifts to be raffled off at a corporate picnic, an expensive meal—where do these fall? Is it ethical for a purchaser to disclose information provided by one supplier to a competitive source?

Familiarity with the profession's and one's organizational ethical requirements is not sufficient. Sound ethical judgment is still required on a day-to-day basis, and constant alertness is required to avoid sliding into unethical behavior.

Various forms of research have been conducted, primarily using hypothetical questions and situations, to test for conformance among both purchasing students and practitioners. That no unanimity existed was not surprising. Unfortunately, even practices that were clearly unacceptable by NAPM's guidelines were deemed acceptable by a significant number of participants in the research.

KEY POINTS

1. Purchase orders and contracts are the two main methods of procurement.

2. Critical commercial content in a purchase order or contract deals with what is to be acquired, how much, for how long, when it is

required, what the price will be, and shipping instructions. Additional terms and conditions cover a host of contingencies and requirements.

3. Blanket orders, standing orders, purchase order draft with order, telephone orders, consignment, and short- and long-term contracts offer a variety of contractual means.

4. Strategic alliances or joint ventures, supplier partners, and preferred suppliers represent different interorganizational relationship options for buyers and sellers.

5. Systems contracting, especially when coupled with EDI and bar coding, offers special opportunities to deal quickly and inexpensively with large numbers of small purchases.

6. Letters of intent preliminary to contract finalization can be either binding or nonbinding.

7. Subcontracting allows a prime contractor to engage other suppliers to perform part of the work required. Purchasers may wish to retain the right to approve subcontractors.

8. Most organizations have policies and procedures that deal with the award of contracts and the notification of unsuccessful bidders.

9. Follow-up ensures that the order will be delivered on time. Expediting represents an attempt to obtain earlier delivery.

10. Proper receipt of goods helps confirm that the purchaser received what was ordered and helps discover discrepancies quickly.

11. Inspection confirms that incoming goods meet the quality standard required. Inspection does not add value. The purchaser has a number of remedies if incoming requirements fail to clear inspection.

12. Invoice checking may reveal discrepancies that need to be resolved before payment is made. It is normal to ensure that the purchase order, waybill, receiving report, and invoice are congruent.

13. FOB terms are critical in the resolution of transportation issues.

14. The purchaser must ensure that the supplier receives payment as agreed in the contract.

15. Record maintenance ensures that an audit trail is available for completed acquisitions.

16. Proper disposal of excess or surplus materials, rejected end products, scrap material, waste, and hazardous waste is becoming increasingly difficult and expensive.

17. Major contract administration, as practised by the government and large organizations, requires contract administrators to ensure that both parties fulfill their obligations and that the contract will be completed on time and within budget.

18. Typical contract administration activities include (1) work control; (2) assuring compliance with contract terms and conditions; (3) assuring financial control; (4) monitoring, auditing and approving contract systems; and (5) other administrative actions.

19. The procurement field is rife with ethical issues. Each organization has to define its own rules and regulations governing the actions of its supply personnel. NAPM, like other professional purchasing associations around the world, provides a code of ethics for purchasing professionals.

CHAPTER 8

LEGAL ISSUES IN PURCHASING AND SUPPLY

This concluding chapter provides an overview of the more commonly encountered legal issues in the purchasing field: the authority and liability of the purchaser, regulations affecting buying and selling, essential contract requirements, special clauses, liability protection, warranties, errors, suppliers in financial difficulties, contract conclusion options, and dispute settlement.

Proper attention to the legality of the contracting process and the use of the appropriate contract provisions will help ensure effective transactions between buyers and sellers. Expert consultation should be sought when the purchaser is unsure about legal issues.

THE LEGAL AUTHORITY OF THE PURCHASER

In most organizations, the purchasing department acquires goods, services, and equipment for use by others (or at the request of others) in the organization. Under the law of agency, an agent is given authority by a principal (person, institution, or corporation) to act on his, her, or its behalf. Thus, a buyer is committed by his or her principal and is expected to act in the best interests of the principal. The buyer has a fiduciary duty based on the trust and confidence of the principal to spend the principal's funds ethically, fairly, and wisely.

Principals can and do limit the authority of the purchaser to commit the organization to suppliers. The job description shows the range of duties for which a buyer is responsible. Delegations of authority or spending limits are used to define upper limits (normally per contract) to which a buyer can commit without obtaining approval. Special agents have limited authority to perform only certain tasks. General agents have broader powers and more latitude to exercise judgment to carry out their duties.

When a board of directors delineates authorized buying limitations in a written document and passes these as a resolution, the agreement is one of general agency. If a general agent engages in unauthorized acts, the principal will be held liable because the general agent acts for the principal. In the absence of an actual agency relationship, apparent authority may be granted by a principal permitting a party to operate in such a way that those dealing with this party have good reason to believe this party is an authorized agent. For example, even though a buyer's authority to purchase may be $25,000, if he or she places an order for $40,000 with a supplier, the organization is bound to the contract.

PERSONAL LIABILITY OF THE PURCHASER

The law of agency does not absolve the purchaser of all personal liability. A purchaser may be held personally liable if he or she:

1.   Makes a false statement concerning authority with intent to deceive or when the misrepresentation has the natural and probable consequence of misleading.

2.   Performs a damaging act without authority, even though believing he or she has such authority.

3.   Performs an illegal act, even on authority from the employer.

4.   Willfully performs an act that results in damage to anyone.

5.   Performs a damaging act outside the scope of authority, even though the act is performed with the intention of rendering the employer a valuable service.

In each of these cases, the supplier ordinarily has no recourse to the company employing the agent because there existed no valid contract between the seller and the purchasing organization. Since such a contract does not exist, the only recourse that the supplier commonly has is to the agent personally.

The principal can still be held liable for contracts made within the apparent scope of the agent's authority but beyond the actual scope because there were limitations on the latter unknown to the seller. Under these circumstances, the agent is probably wrong and is, of course, answerable to the principal. Buyers may also be answerable to the seller on the ground of deceit, on the charge that they are the real contracting parties, or for breach of the warranty that they were authorized to make for the principal.

Moreover, suits have been brought by sellers against purchasing managers when it was discovered that the latter's principal was for some reason unable to pay the account. For example, such conditions have arisen when (1) the employer became insolvent or bankrupt; (2) the employer endeavored to avoid the legal obligations to accept and pay for merchandise purchased by the purchasing manager; or (3) the employer became involved in litigation with the seller, whose lawyers decided that the contract price could be collected from the purchasing manager personally.

REGULATIONS AFFECTING PROCUREMENT

Many laws and regulations affect the procurement area. Some deal with the bidding process, others with prices and contracts, and others with national and international trade and the environment. An overview of these regulations follows.

Regulations Affecting the Bidding Process

Verbal versus Written Quotes

Under the Uniform Commercial Code (UCC), verbal contracts are enforceable up to a maximum of $500. Typically, the request is made by telephone; the supplier is given an identifying order number and is advised that no confirming written document will be forthcoming.

As summarized in Ritterskamp's *Purchasing Manager's Desk Book of Purchasing Law*, Section 2-201 of the UCC contains the following:

Subsection (1). An oral contract for the sale of goods for the price of $500 or more is not enforceable in a court of law unless there is some writing indicating that a contract was made between the two parties. The writing must be signed by the party who would be the defendant if the contract was to be the basis of a lawsuit.

The Statute of Frauds is a law designed to prevent frauds and perjury by requiring that contracts for amounts over $500 should be in writing. This area of contract law is designed to protect people from false testimony. Evidence in writing, of course, reduces the possibility of such an occurrence. The Doctrine of Promissory Estoppel holds that if one party makes a promise, even orally, he or she cannot renege on that promise. Accordingly, while an unwritten contract for more than $500 might be unenforceable under the Statute of Frauds, it might be enforced under the Doctrine of Estoppel if sufficient evidence of the agreement is available.

An oral contract for services is also enforceable in court without writing if the services can be performed within one year from the date of the contract.

Fax. As a general rule, documents transmitted by facsimile are considered as valid as the original documents. Obviously, sealed bids cannot be transmitted by fax. However, if an RFQ form has terms and conditions on the reverse side that are not generally transmitted with the faxed copy of the document, it is important to establish beforehand with suppliers that they acknowledge and accept the buyer's standard terms and conditions as part of business transactions conducted via fax.

Regulations Influencing Prices and Contracts

Many federal, state, and local regulations affect how business is conducted and apply in varying degrees to purchasing activities. Here are the most frequently cited:

1. The Sherman Anti-Trust Act prohibits contracts, combinations, and conspiracies that result in a restraint of trade. Whether a business

arrangement or agreement can be construed as a conspiracy in restraint of trade depends on the facts and circumstances involved. In most cases, the courts will decide if the facts show a reasonable and understandable business arrangement or a conspiratorial arrangement designed to impair the free market system. In Canada, the equivalent legislation is covered by the Combines Investigation Act.

2.   The Clayton Act deals with trade practices like tie-in arrangements, full-line forcing, and exclusive dealing, which are unlawful where the effect may be to lessen competition substantially or to create a monopoly. Tie-in sales occur when a seller requires the purchaser to buy one product in order to purchase another product. They are expressly forbidden, especially when a manufacturer makes pricing for two items so unreasonably attractive relative to individual prices that the buyer must buy the total package. The Clayton Act also prohibits mergers and acquisitions that may harm competition or create a monopoly.

3.   The Robinson-Patman Act prohibits price discrimination for goods of like grade and quality where the end result may be to lessen competition substantially or create a monopoly. Discrimination in services and promotions that is connected with the sale of a product can be unlawful even where there is no injury to competition. Specific provisions of the act that apply to purchasing are:

   a.   It prohibits direct and indirect price discrimination where those price differences substantially lessen competition.
   b.   It allows the seller to offer lower prices if this is done in good faith to meet an equally low price of a competitor.
   c.   It prohibits a seller from paying a commission to a buyer and prohibits a buyer from accepting it.
   d.   It requires that payments for any facilities or services furnished by a buyer to a seller be equally available to all other customers.
   e.   It requires that facilities or services furnished to a buyer by the seller be equally available to all of the seller's other customers.

f. It prohibits a buyer from knowingly inducing or receiving a price discrimination.

Price discrimination is permitted under the law when necessary to meet the price of a competitor who is selling goods of like grade and quality. Discrimination is also permitted where there is cost justification for the differences. Typically, these differences apply to cost of production and distribution and to deteriorating, perishable, or seasonal goods. Thus, quantity discounts must be offered equally to all buyers and must be traceable to the economics of quantity production and distribution. Robinson-Patman does not prohibit price reductions, only discriminatory and preferential pricing. Additionally, if goods are made specifically to a buyer's specifications, a seller may offer a lower price to one buyer than another because the products are not of like grade.

4.   The Federal Trade Commission (FTC) Act created the FTC in 1914 and gave it the power to determine the meaning of "restraint of trade." The FTC is charged with uncovering unfair methods of competition and unfair or deceptive acts in commerce. All proposed corporate mergers must withstand the FTC test for unfair competition.

5.   The Davis-Bacon Act and related laws regulate minimum wages for public works employees.

6.   Article 2 (Sales) of the Uniform Commercial Code (UCC), completed in 1961, and its several revisions, have been generally adopted by all states in the United States except Louisiana. The provisions protect parties to commercial transactions involving the sale of goods to the end that the aggrieved party may be put in as good a position as if the other party had fully performed.

7.   The Prompt Payment Act provides requirements to government procurement offices that ensure that federal contractors supplying goods and services are paid on time.

8. The Service Contract Act holds that the Wage and Hour Division of the Employment Standards Administration is responsible for (1) predetermination of prevailing wage rates for federal construction contracts and (2) a continuing program for determining wage rates under the Service Contract Act.

9. The Walsh-Healey Public Contracts Act, responsible for federal procurement reforms, contains three major provisions:

   a. *Revolving door*—Federal procurement officers must excuse themselves from dealing with contractors if they have had prospective employment discussions with them. Furthermore, if a federal procurement officer is the prime negotiator for a contract, that person cannot work for the contractor for a minimum of two years after conclusion of negotiations.

   b. *Allowable costs*—Contractors supplying goods and services to the Government are barred by law from receiving reimbursement for costs such as lobbying, entertainment, or contract-related legal fees.

   c. *Multiple sourcing*—The Secretary of Defense is required to plan for the use of multiple sources in both development and production of major weapon systems. Deviation from this requirement can be granted on the grounds of national security.

10. The Small Business Act ensures that small businesses receive a fair share of the government's procurement dollar. There are two types of procurement from small business: (1) those assigned solely to small business and (2) those divided between small and large business. Before RFQ, RFI, or RFP packages are issued, the government buyer determines if the purchase has small business application. Bid packages are issued to suppliers accordingly. In the case where the business is to be divided, small businesses are given an opportunity to match the bids submitted by large businesses, since the government does not pay a premium price to source from small business.

11. The Buy America Act requires that government purchases for public use consist only of raw materials mined or produced in the United States, or manufactured items which are made in the United States from materials or items mined, produced, or manufactured in the United States Exceptions to this requirement are: (1) if the items are not available in the United States in commercial quantities of good quality, (2) if the cost of the domestic items is unreasonable (generally, if the cost of the domestic items is over 6 percent more than the cost of comparable foreign items), or (3) if the head of the department decides that it is in the public interest to waive the requirement.

12. The Trade Act, passed in 1988, contains laws relating to the following topics: trade policy and negotiations, unfair foreign trade practices, import relief, trade adjustment assistance, foreign subsidies and dumping, intellectual property rights, tariffs and customs, export promotion, international finance, agriculture, foreign corrupt practices, and education and training.

13. The North American Free Trade Agreement's major provisions covering the United States, Canada, and Mexico define the regulations for the following: tariffs and customs, import and export restrictions, industry-specific provisions, emergency actions, and dispute settlement.

14. The Foreign Corrupt Practices Act prohibits the bribing of foreign officials. It prohibits payments to foreign officials that are meant to influence the award of business. It defends payments that are made in good faith or reasonable expenditures for production, promotion, or contract performance.

15. The Federal Acquisition Regulation (FAR) is the body of primary regulations used by all federal executive agencies in the acquisition of supplies and services with appropriated funds. The FAR is a compilation of all the laws and policies governing the federal procurement process.

This act also created the Federal Acquisition Regulatory Council, which consists of the head of the Office of Federal Procurement Policy and the heads of three major procuring agencies: the Defense Department, the General Services Administration, and NASA. The council directs and coordinates governmentwide procurement policy.

16. The False Claims Act provides for the recovery of damages and remedies upon proof to the government of loss sustained through fraud in the award or performance of government contracts.

17. The Truth in Negotiations Act requires that, for noncompetitive contracts in excess of $100,000, contractors must furnish the government buyer with complete, accurate, and current data on relevant costs. It also requires that suppliers certify that all data furnished are current, accurate, and complete.

18. The Armed Services Procurement Act, passed in 1947, established workable procedures for procurement in periods of national emergency. It also recognizes that negotiated procurement is a required method of purchasing in peacetime as well as wartime.

19. The Competition in Contracting Act (CICA), passed in 1984, promotes the use of full and open competition in the procurement of property and services by the federal government. Prior to this act, the two basic government procedures were formal advertising (sealed bids and award to lowest bidder without discussion) and negotiation.

20. The Model Procurement Code for State and Local Governments (MPC), completed in 1978, provides statutory guidance for the procurement of supplies, services, and construction by state and local governments and administrative and judicial remedies to resolve controversies relating to public contracts at the state and local government level.

21.   The United Nations Convention for the International Sale of Goods (CISG) was ratified by the US Senate in 1986; as of this writing, it has been ratified by 34 nations. Additional nations will join the convention as their legislatures act. The convention is now part of US law. Purchasing officers and legal counsel must observe the terms of the convention unless the parties have agreed up front to waive the convention and apply the law of the country in which either the buyer or the seller is located. Like the UCC, the CISG is not intended to cover the sale of services. However, it does recognize that some contracts could cover a mixture of goods and services. The preponderant part of such contracts will determine applicability.

22.   The Resource Conservation and Recovery Act (RCRA), passed in 1976, was the cornerstone hazardous waste control law, although it was not implemented by the Environmental Protection Agency (EPA) until 1980. The RCRA identifies which industrial wastes are corrosive, toxic, ignitable, or chemically reactive. The EPA regulates how these potentially dangerous items are to be treated, disposed of, or stored.

23.   The Toxic Substances Control Act, passed in 1976, regulates procedures used in the manufacture, use, and distribution of chemicals. Included are polychlorinated biphenyls (PCBs), which the 1979 EPA regulations specify must be disposed of in chemical waste landfills.

24.   The Comprehensive Environmental Response, Compensation, and Liability Act, passed in 1980, including Superfund Amendments and the Reauthorization Act of 1986, is known as SARA or Superfund. It gives the EPA authority to order the cleanup of solid waste sites and to recover the costs, if necessary, from those responsible for the pollution damage.

25.   In addition to federal, state, and local regulations that may affect the purchasing professional, the federal courts and the FTC have added to the present laws through their interpretation and

application of these various acts. This case law expands the general statements of the laws cited above.

PURCHASE ORDERS AND CONTRACTS

When a buyer agrees to purchase from a seller and the seller agrees to provide what the buyer requests, a contractual agreement has been reached. This contract is evidenced and defined by both the language and actions of both buyer and seller. The agreement must be legal to be enforceable, may cover almost any good or service and must meet the basic obligations of good faith, diligence, and reasonableness. Contracts may be oral or written, and as long as both parties perform as agreed, satisfaction for both sides should be the result. Purchasers and sellers alike can avoid many potential pitfalls by ensuring that the contract is legally binding, complete in its coverage, and committed to by both parties.

Contract Elements

A proper contract must meet five essential elements: (1) offer and acceptance, (2) consideration, (3) competent parties, (4) legality of purpose, and (5) genuine assent.

1. Offer and Acceptance

A contract can exist only if an offer has been made by one party and the other party has unconditionally accepted that offer. Interestingly enough, price quotations and advertisements are not construed as offers, and neither are discussions about delivery, quality, quantity, and price. All of these are construed as activities leading up to contract agreement. In the buying-selling context, therefore, it is normally the purchase order that is construed to be the offer and the supplier's acknowledgment thereof the acceptance.

Acknowledgment. Normally, a supplier's acknowledgment is a standard form sent to the purchaser to acknowledge receipt of the purchase order; it generally implies acceptance of this order. In the absence of a formal acknowledgment, the Uniform Commercial Code

(UCC) states that if parties deal as though they have a contract, then they have one. For example, a purchaser requests a supplier to perform a maintenance task on a specific date on a specific piece of equipment, but receives no written acknowledgment. If the supplier's personnel arrive on the appointed day and are granted access to the premises and the equipment, it is presumed a contract is in effect.

In international buying, however, under CISG Article 18 (3) a contract must contain specific agreement that acceptance may be indicated by delivery of goods.

Counteroffer. A counteroffer is an offer to enter into a transaction on terms different from those originally proposed. A counteroffer represents a new and different offer and terminates the original offer, even though a substantial number of terms and conditions of the original offer have not been changed. Therefore, if a supplier's acknowledgment form stipulates terms and conditions that are different from those on the original purchase order, it is a counteroffer and not an unconditional acceptance. In standard commercial practice, many buyers tacitly ignore the differences between forms and assume the supplier has accepted their order as stated. Section 2-207 of the UCC states that

> even a sales acknowledgment containing terms and conditions in addition to or different from those in the purchase order will serve as acceptance and the additional terms will become a part of the contract, unless:
>
> 1. Acceptance is expressly conditioned on buyer assent to the additional or different terms.
> 2. Acceptance is expressly limited to the specific terms of the offer.
> 3. They materially alter the contract.
> 4. The buyer's objection to the additional terms is given within 10 days.

In any of the above four cases, the sales acknowledgment will constitute a counteroffer to which the buyer must then respond within 10 days. This back and forth "jockeying" for position is referred to as the "battle of the forms."

Even if the writings between parties do not in and of themselves establish a contract, a contract may be established if the parties conduct themselves in a way that recognizes such a contract. In such cases, the terms of the agreement will consist of those on which the writings of the parties agree. Conflicting terms in the PO and sales forms will essentially cancel themselves out, and any remaining gap will fall under the code.

When a purchaser does not receive an acknowledgment, the contract is considered a unilateral one in which the actions of the other party constitute acceptance. A bilateral contract contains a promise to buy and a promise to sell under the same terms of agreement.

2. Consideration

Consideration refers to the exchange of value for value to make the promise binding. In consumer purchases this might entail a deposit or down payment. In commercial acquisition such up-front payments are not commonly used, except in situations where good faith needs to be evidenced or a supplier is providing for a unique purchaser need that might require substantial supplier cash to produce or develop. In the purchase of standard goods and services on a repetitive basis, such advances or deposits are seldom used. Aside from money, consideration may also be satisfied by giving up an existing right such as goods, an act or service, or an exclusive dealing promise.

3. Competent Parties

Both parties must be in a position to enter into a contract and be capable of doing so.

4. Legality of Purpose

The contract must be consistent with federal, state, or provincial constitutions and cannot violate legal statutes or public policy. Moreover, aside from the contract, the consideration offered must be legal. Illegal drugs cannot be offered as payment. Agreement to restrain trade is not legal and therefore unenforceable as a contract.

5. Genuine Assent

A contract is unenforceable if it is made under duress or as a result of one party's undue influence over the other. A seller cannot force a purchaser to purchase product A, which the purchaser does not need, as a condition to providing product B, which the purchaser needs desperately and for which no substitute is available. A supplier's representative cannot pull a gun on a buyer and insist that the buyer sign a purchase order. Fraud and misrepresentation also invalidate a contract. A supplier has a legitimate reason for not observing a contract if it comes to light that a purchaser deliberately provided requirements estimates higher than expected to induce the supplier to quote a lower price.

These five elements are essential to a contract from a legal perspective. To be meaningful and effective in a commercial sense requires substantially more in terms of contract nature and content to assure satisfaction for both buyer and seller. Therefore, the following discussion covers types of contracts and their content.

Types of Contracts

Contracts may be oral or written. The following sections will discuss both types.

Oral versus Written Contracts

A contract may be made orally or in writing provided both parties have abided by the five elements of contracts discussed earlier. Under the UCC, contracts in excess of $500 are not enforceable unless written evidence can be provided that they exist, or if:

1.　The goods are made specifically to the buyer's order and cannot readily be sold to others (such as a printing order for forms with the organization's logo on it).

2.　If both parties behave as if the order existed. (The organization sends artwork with the logo to the printer and the printer runs 100,000 copies and delivers on the date promised.)

3.　If the parties have always done business on an oral basis.

Oral orders have traditionally been frowned upon by prudent buyers. Oral orders have been used for low dollar value purchases made from approved and reliable sources to meet important or emergency deadlines. Standard practice has been to validate oral orders with a confirming purchase order. These days, with an increased emphasis on value-added activities, empowerment, lead time reduction, and downsizing, confirmation purchase orders have become less common. For one thing, the cost of the confirmation order may exceed the value of the purchase. Alternatives may include release against a systems contract by a user, oral orders recorded on audiotape, an oral order memo pad, or electronically recorded orders. Credit cards have also become a popular alternative because they provide a record of the purchase and facilitate billing and payment practices.

Avoiding the need for confirmation orders by policies and procedures that incorporate oral orders, oral releases, or suitable alternatives is better purchasing practice.

LEGAL ISSUES AND THE CONTRACT

The purpose of any contract is to commit both parties to perform according to their agreement. The fundamental proposition in a buying-selling contract is that the buyer wants to purchase requirements and the seller wants to provide them. A well-defined and well-structured contract will permit both parties to focus on their respective responsibilities without having to renegotiate, seek clarification, or cover up for omissions or mistakes. A complete contract may cover a wide range of potential issues. The following list is intended to provide an overview of the points that may be covered in the contract beyond the standard coverage of what is required, how much, when, what price will be paid, and what delivery arrangements will be.

Both the applicability of certain clauses and their appropriate wording should be checked with legal counsel before contract completion. Many purchasing organizations maintain a file of standard clauses for potential inclusion in contracts. In repetitive situations this can speed contract formation and avoid unnecessary legal consultation. In nonrepetitive situations or very large contracts it is prudent to ask for the assistance of legal counsel before contract signing to ensure that the appropriate contract areas have been covered and that the standard clauses selected are sufficient.

Special Clauses and Provisions

A purchaser's decisions need to be strategically, economically, ethically, and legally sound. On the legal side, prevention of future legal difficulties is highly preferable over short-term expediency. The following list identifies common contractual areas:

1. *Strategic or large contracts.* These require special attention because of their potential impact on organizational success.

2. *Strategic alliances or preferred supplier or purchaser-supplier partnering arrangements.* These require special attention to the buyer-supplier relationship issues beyond the standard contractual terms.

3. *Single or sole source contracts.* When a buyer places 100 percent of a requirement with one supplier, a single source is created. Issues of quality and supply assurance and price provisions are of extreme importance in such contracts.

4. *Patents.* Patents are issued after due proceedings on application to the Commissioner of Patents. US patents grant their owners, for a limited period, the right to exclude others from making, leasing, using, or selling patented inventions throughout the United States. The patent owner may license others under the patent. Patents may cover a process, machine, method of manufacture, composition of matter, or any new or useful improvement.

    A buyer's firm may be sued for patent infringement by virtue of its use of a patented item even though the item was purchased in good faith. If valid, the suit could result in payment of damages and an injunction against continued use of the item. The UCC contains a patent warranty making sellers responsible for any damages incurred by a buyer, but does not require such payment until the case is decided. Since patent suits may be very costly, a buyer should first ensure that his or her purchasing form is asserted and, secondly, ensure that a well-designed indemnification clause gives him or her rights against the seller, including (1) a provision for periodic reimbursement of costs as the suit progresses; (2) the right to require the seller to defend the suit; and (3) the right to have the buyer's lawyers involved in the suit.

If conflicting terms cancel each other out and such a suit falls back under the code, only the patent warranty will apply. One way to avoid this is to have the seller sign the PO acknowledgment or some other form containing buyer terms.

The terms "patent pending" and "patent applied for" mean that the patent is before the Patent Office for consideration. There can be no infringement of a patent sought by the applicant. While many applications are rejected, a patent-pending application should be a warning to a buyer. In seeking a second source for a patented item, a buyer should never purchase from a nonlicensed source on a "copy or equal" basis.

Purchasing personnel should also be instructed to gain rights for their organization whenever possible. An invention can become a valuable asset. Accordingly, the purchasing department should consider each purchase thoroughly prior to the award of a contract in order to decide whether rights to a by-product invention are a subject to be negotiated. The purchaser should receive all rights, title, or interest in any invention or process that was discovered by a supplier with the use of the purchaser's funds.

5. *Copyrights.* Copyrights are rights granted by law similar to those of patents, except that they grant monopolistic rights to an author, artist, or composer for the publication, production, or sale of rights to a literary, dramatic, musical, or artistic work. Most of the precautions cited above to protect against patent infringement also apply to copyrights. As information services increase in importance, buyers must be careful not to violate copyrights on published reports, software manuals, and so forth.

6. *Limitation of liability.* Limitation of liability relates to programs where sellers devise ways to evade or reduce their liability under warranties. Two ways sellers attempt to do this are the parole evidence rule and the merger or entirety clause. The *parole evidence rule* states that the writing involved in the contract

becomes the sole evidence for the contract. Applying this rule means that any statements about product quality made by the seller's agent during negotiations and not put in writing would not be valid.

In using the merger or entirety clause, the seller states in his or her acknowledgment that the agreement constitutes the final written expression of all terms and that any representations, promises, warranties, or oral statements by the seller's agent that differ from the agreement are not in effect.

7.   *Waiver of consequential damages.* Consequential damages are damages beyond what are known as *incidental damages* (such as inspection, transportation, and replacement expenses). Consequential damages relate to lost profits, lost sales, and injury to property or people. In a waiver clause, the seller states that the buyer's sole remedy is limited to the repair and/or replacement of parts deemed defective, and that no consequential damages for lost profits, lost sales, or injury to property or people will apply.

8.   *Exception to approved terms and conditions.* In cases when a seller's terms differ from the buyer's and a dispute arises, whose terms apply? The UCC guidelines state that if the seller's form appears to a reasonable and prudent buyer to be an acceptance to a given offer, it will probably operate as an acceptance, even though there are additional clauses added. Any material change affecting either the buyer's rights or the seller's performance under the contract will not become part of the contract.

9.   *Suspension.* Suspension clauses protect the buyer against unforeseen funding problems, delays on the part of internal personnel, or changes in management thinking. A suspension clause states that all or part of the work under the purchase contract may be suspended by the buyer at his or her convenience. Payment for expenses incurred up to the time of suspension are reimbursable if properly documented.

In the public sector, suspension is a term applied when a supplier is found to be in nonconformance regarding quality or delivery performance. The suspended supplier has a certain time period to correct his or her performance.

10. *Collusive offers.* Collusion is a secret agreement and cooperation for a fraudulent or deceitful purpose. Federal antitrust laws are designed to prevent collusive activities on the part of sellers. Buyers may need to seek the advice of counsel in cases where collusion is suspected.

11. *Force-majeure.* Force-majeure translates from French as "a major or superior force." In contracts, it refers to major (and often uncontrollable) events that excuse a party in whole or in part from performance of obligations. For example, a party may be excused from performance in case of a fire, a war, or a strike. The term is colloquially referred to as the "act of God" clause, and is normally included in standard purchase order forms to protect both buyer and seller. If the seller or buyer cannot perform because of some intervening force that was not foreseeable when the contract was made, performance of the contract will be excused without liability.

12. *Comity.* Comity means that the courts in one state recognize or accept the laws of another state. For example, it may be possible for a buyer or seller to sue the other party in state A on a contract that was signed in state B. When the court in state A uses the law of state B in interpreting the contract, this denotes comity. The actual state laws that apply are negotiable if buyer and seller are located in different states. With assistance of legal counsel, a buyer can determine which state's laws favor the buyer and which favor the seller and can attempt to negotiate the best alternative.

13. *Reservation of rights.* The buyer reserves all rights to contract performance in accordance with stated terms and conditions. A terms clause prohibits sellers from injecting any waivers of warranty or terms change without the consent of the buyer. Typical wording in a contract would be, "This purchase order, together with any attached drawings or specifications incorporated herein, sets forth the entire terms and conditions, and no other terms and

conditions in any document, acceptance, or acknowledgment shall be effective or binding unless expressly agreed to in writing by the buyer."

14.   *Estoppel.* Estoppel is the legal principle that prevents a person from asserting a position that is inconsistent with his or her prior conduct if injustice would thereby result to a person who has changed position in justifiable reliance upon that conduct.

15.   *Hazardous and regulated materials.* Several federal and state laws apply to hazardous or regulated materials. In 1976 the Toxic Substance and Control Act was passed to control the manufacture, distribution, and sale of chemicals. The EPA maintains a list of all existing chemicals considered toxic and can limit the use of chemicals shown to present unreasonable risk to health or the environment.

   The Resource Conservation and Recovery Act, also passed in 1976, established a plan for the transportation and disposal of hazardous waste. The EPA uses four criteria for identifying such waste: (1) ignitability; (2) corrosivity; (3) reactivity; and 4) toxicity. Purchasers will typically need to consult with legal counsel on the interpretation of these and similar environmental laws.

16.   *Restraint of trade/antitrust.* Buyers should keep in mind that the purpose behind all antitrust legislation is to protect the US free enterprise system. The four primary laws dealing with antitrust are the Sherman, Clayton, Federal Trade Commission, and Robinson-Patman Acts. Buyers frequently need to seek counsel for advice on these laws.

17.   *Assignability provisions.* Assignment means a transfer of one's rights under a contract or the transfer of both rights and duties. In their standard terms and conditions, many buyers ask the seller to secure buyer approval prior to assigning its performance under the contract to another party. Assigning all or part of the work without the buyer's approval would be cause for termination of the

contract. A typical statement in a contract is, "The seller shall neither assign its performance under the purchase order or any part thereof, nor delegate any obligations hereunder without the prior written consent of the buyer."

18. *Source code escrow accounts.* On installment sales contracts the buyer may request an escrow account. The escrow agent (third party) physically stores the property and has title to it. Upon buyer performance to the contract terms and conditions, the title is transferred to the buyer. Otherwise, the buyer bears a risk that the seller may not convey title upon the buyer's satisfying contract requirements.

19. *Domestic versus international legal relationships.* More terms are becoming standardized between nations as worldwide market dealings become more common for both buyers and sellers. "Uniform customs and practice for documentary credit" apply to the handling of letters of credit. "Inco terms" establish standards of responsibility and title for international transportation. But many other terms and conditions are less standardized. The courts of certain foreign nations recognize or accept the laws of another nation (comity). Knowledge of comity is especially important when placing purchase orders with overseas suppliers. When dealing with foreign suppliers, the issue of court jurisdiction is frequently a major stumbling block during negotiations. Determining with legal counsel whether the laws of the supplier's country recognize the laws of the United States may resolve this negotiating issue. Often buyers prefer to deal with offshore suppliers who have US based assets. If a dispute occurs, these assets can be tied up in court and provide a basis for compensation.

20. *Protests.* Suppliers may protest awards if they feel unfairly treated. In the public sector, the process is more formal. Typically, an administrative hearing board will listen to the supplier's complaint and decide if an appropriate remedy is required. During the hearing process, the award is held in abeyance until the supplier's protest is ruled upon. Public sector purchasers in particular are advised to consult with legal advisers on how to handle a bid protest properly.

21.   *Claims.* Claims include any right to payment or to receive any equitable remedy such as specific performance of a contract. Most often, claims arise in conjunction with a bankruptcy, when the debtor files a list of creditors and the court sends a notice of bankruptcy proceedings to the creditors. Creditors must then file proof of claim to participate in the assets of the estate.

22.   *Insurance.* Sound purchasing practice dictates that sellers or contractors who come onto the buyer's site have adequate insurance coverage for damage and personal injury to themselves and others. Recommended coverage includes commercial general liability, automobile insurance, worker's compensation, and employer's liability insurance to appropriate limits, designed to fit potential hazards. Subcontractors should be similarly insured, and the purchasing manager should make the purchase contract conditional upon adequate coverage and ensure that it is in place before issuing a contract. Most buyers specify the minimum amount of insurance required within the contract; for example, "The seller shall carry insurance protecting the buyer in not less than minimum limits of property damage of $1,000,000 and public liability of $500,000/$1,000,000."

23.   *Insurance and indemnification.* The terms "hold harmless," "defend," and "indemnity" commonly refer to indemnification clauses used to protect a buyer or seller from loss or damage. Each word has a subtle difference. *Indemnity* is defined as legal exemption from penalties or liabilities incurred by one's actions; it normally pertains to financial or monetary loss. The term *defend* imparts an obligation to one party of a contract to defend the other from any legal action and to incur the cost of such legal action. *Hold harmless* (or "save harmless") is a legal term that normally pertains to consequential damages or economic injury. Such clauses are standard boilerplate in most purchase order forms. They are also used by buyers in construction contracts to protect against claims served on contractors.

Contract Liabilities

Contracts require that both parties be liable for their actions. A prudent purchaser attempts to limit the liability exposure of his or her organization while at the same time protecting its rights to remedy in the event of future conflict.

The intent of the UCC is not to penalize one who defaults under a contract but to the aggrieved party for direct damages caused by the breach, together with any incidental or consequential damages resulting from the breach. That is, the intent is to make the aggrieved party whole again. However, before he or she can receive damages, a buyer must specifically prove them. Further, the seller is allowed under the code to limit its liability for incidental and consequential damages and will generally do so.

Purchase order terms and conditions should always contain clauses that disclaim or limit a buyer's responsibility for damages resulting from seller violation of existing law. However, the language typically used on a sales form attempts to limit seller liability or to place that liability elsewhere, and buyers must defend against those efforts. Specific concerns include, but are not limited to, the following:

1. The Consumer Product Safety Act protects consumers against unsafe products. It requires that known or potential hazards in consumer goods be reported to the Consumer Products Safety Commission and that unsafe goods be recalled. It covers manufactured goods and goods purchased for resale, including components. Purchase forms should contain a clause placing liability with the manufacturer indemnifying the buying firm from costs associated with recalls or defects. It is also good practice to require suppliers to notify the buyer of known or potential product hazards.

2. In regard to product liability the law generally states that people injured by or because of products have a right to sue and that each party in that product's distribution chain is jointly and severally liable. However, any party in the chain who did not contribute to the problem and who is not rightly responsible has a right of

indemnification against those who are. Furthermore, the parties to a contract can make prior agreement as to who will bear responsibility for personal injury.

As the seller's form is likely to contain product liability disclaimers, purchasing forms should contain proper indemnification clauses so as to effectively preserve the buyer's rights.

3.   Patent liability and insurance have been discussed earlier.

Warranties

A warranty is a promise made by the seller that is legally enforceable as long as it is included in the contract. A buyer is free under the law to bargain for broad, strong warranties or to accept a seller's total and complete disclaimer of warranties. Warranties fall into two categories: express and implied.

Express Warranties. An express warranty is expressed orally or in writing and can include just about any statement or representation a seller makes about its product or service. In addition to specifications as detailed on the face of a purchase order, express warranties may include advertisements, catalog descriptions, photos, proposals, samples, and even oral claims by the seller, provided they are included in the contract. A purchase order form nearly always contains a clause establishing that it constitutes the full and complete agreement between the parties. Therefore, to include such expressions in the contract (and thereby make them warranties) they must be referenced in the contract writing.

Implied Warranties. Implied warranties are provided by law. That is, a buyer does not have to list them specifically on the face of a contract in order to have them apply. Implied warranties include the following types:

1.   *Warranty of title and authority to sell.* It is implicit in contracts between merchants that the seller has legal ownership of the goods being sold, that they are not subject to security interests, liens, or

any other encumbrance not known to the buyer at the time of contract formation, and that the seller has a right to sell or otherwise transfer the goods accordingly.

2. *Implied warranty of merchantability.* Unless the warranty is excluded or modified, it is implicit in a contract for goods that they will be merchantable. This means that the goods must be of fair average quality, must pass without objection within the trade, and must be fit for the general purpose for which such goods are normally used.

3. *Fitness for an intended purpose.* When a contract is formed, if a seller is aware of the purpose for which goods are being purchased and that the buyer is relying on the seller's expertise to furnish a suitable product, it is implicit that the seller warrants the goods to be suitable for that purpose.

Note that the UCC allows sellers, by conspicuous use of specific language, to disclaim the implied warranties of merchantability and fitness for a particular purpose and to limit express warranties to repair or replacement. Virtually every sales acknowledgment form contains such language. It is therefore important that buyers use forms with well-designed terms and conditions to cancel out seller disclaimers, after which remaining gaps will be filled by the UCC.

Failure of Essential Purpose. The first paragraph of section 2-719 of the UCC gives sellers the right to limit warranties to their choice of repair or replacement options. This section is typically the source of seller arguments to support a repair position. However, the code still upholds that at least minimum remedies must be available and, while a repair or replacement clause may be reasonable in many cases, in others it may not give a buyer the benefit of the contract originally agreed to.

The purpose of a remedy is to give the buyer goods that conform to the contract within a reasonable time after a defect is discovered and the seller is notified. Section 2-719 of the UCC goes on to provide that where an otherwise fair and reasonable clause operates because of circumstances to deprive a party of the substantial value of the bargain, it fails in its essential purpose. Under these circumstances, such a clause would be thrown out and would give way to general remedy provisions of the code. This would not alter any effective limitations on consequential damages.

Effective Date. Buyers should be concerned with the effective date on which a warranty period begins. For example, it is not uncommon for considerable time to elapse between the receipt of capital equipment and its start-up. If the warranty begins upon receipt, the effective result is a shortened warranty period. Whether a warranty period begins upon receipt of goods, upon start-up, after start-up and "debugging," or at any other time is a negotiable matter, and the parties are free to agree and contract as they wish.

Suppliers in Financial Difficulties

Suppliers in financial difficulties represent several major challenges for purchasers. A whole set of problems may arise when the cash-poor supplier attempts to cope with its financial difficulties. Stretching accounts payable and nonpayment of suppliers may result in supply difficulties or switching to less desirable sources. Trying to cut on quality, service, support staff, wages, salaries, and overhead may adversely affect quality, delivery, quantity, and services.

However, the purchaser can assist in several ways. On condition that the supplier lives up to its commitments, the purchaser can make milestone or progressive payments, supply materials owned by the purchaser, help persuade hesitant suppliers to supply, or provide any other assistance necessary to ensure proper supplier performance. Sometimes the purchaser may, under a separate agreement, take early possession or control of materials without a transfer of ownership or title.

Supplier bankruptcy is particularly serious when (1) purchase contracts are in progress and (2) the seller possesses property of the buyer. Bankruptcy laws permit a supplier who applies for and obtains protection of the courts to look at all executory contracts and decide which it will perform. Since bankruptcy laws are federal, they supersede state laws that govern PO terms reserving the buyer's right to cancel in the event of supplier bankruptcy. Furthermore, bankruptcy laws will not permit a buyer to cancel based on his or her request and subsequent failure to receive adequate assurance of seller performance, even if the PO contains such a clause, except when the seller is already in default.

If the supplier is already in default, then the buyer can request adequate assurance of performance, and the seller will have to (1) cure the default; (2) compensate buyer losses resulting from the default; and (3) provide adequate assurance of performance.

Speed is of the essence in bankruptcy situations. In the first place, the purchaser's organization needs goods and services that may be at best delayed and at worst not forthcoming. Thus, the decision whether to continue with the bankrupt supplier or scramble to obtain another source needs to be resolved quickly. Second, strange happenings often occur at times of bankruptcy, and the purchaser should act quickly to protect the interests of the buying organization.

When the seller possesses property of the buyer, recovery must be handled through the courts. In such a case, the buyer has legal right to its property and is essentially in the position of a secured creditor. Sound practice dictates that such goods be properly marked and that all documentation covering them specifically identify the owner. If the goods are not returned, the buyer can sue, but it might not be worth the effort if the supplier is bankrupt. One safeguard, if appropriate, is for the buyer to file a formal security interest in the property as part of the original agreement.

PO forms should contain the usual terms and conditions dealing with insolvent or bankrupt suppliers, even though they may be superseded by federal law, because they may provide some leverage in the lineup for payment. Also, they could have value if a supplier becomes insolvent and ceases to do business without seeking protection of the courts.

Obviously, the best protection against having to deal with bankrupt suppliers is to be vigilant about the supplier's financial condition before a contract is awarded.

Errors and Omissions

Buyers or sellers may make errors or commit omissions without the intention of doing so. Of course, due vigilance is required of both sides to avoid such situations, but given the large number of transactions that take place in the acquisition field, errors and omissions do occur. The "Law of Mistake" is not clearly defined, because the UCC has no provisions to cover honest mistakes in contract documents. We must therefore turn to traditional law. The courts will not consider relief unless the mistake is material, and then the following general rules will apply:

1. The courts will attempt to be fair with both parties. If a mistake is made that does not cause damage to the other party, relief may be granted. However, the court will not grant relief to the mistaken party if the other party is damaged through reliance on an honest mistake.

2. Relief will usually be granted where one party induced the other to make a mistake or where the mistake is so obvious that it reasonably should have been recognized as such by the party attempting to take advantage of it.

Of course, it is preferable to resolve issues arising from errors and omissions without having to resort to legal means. Goodwill and common sense on the part of both parties should normally prevail.

Typical Purchasing Errors and Omissions
Typical purchasing errors and omissions tend to fall into description, quality, quantity, and price categories.

Obviously, if a description is so erroneous that no one can identify what is to be acquired, clarification between the two parties will be required. If the error is such that (1) a purchase order described incorrect goods, (2) the seller had no reason to recognize the error, (3) the goods were delivered, and (4) the matter cannot be resolved through mutual common sense, the matter will likely fall under the "Law of Mistake," which, with exceptions, will generally not provide relief for unilateral mistakes. It is good practice, therefore, to ensure that purchase orders contain correct descriptions.

The use of an incorrect catalog or part number or the omission of key tolerance or finishing information can easily lead to significant problems. The buyer must check information provided by others in the organization for accuracy and completeness to minimize or eliminate errors and omissions.

Quality descriptions may fail to include tolerance or test specifications or contain the wrong information. Quantities requested may be too low or too high with obvious consequences. Delivery information may be incorrect or too vague (such as ASAP). Delivery instructions may be omitted. Pricing errors may result in incorrect billing, late payment, or more serious consequences.

All of these purchaser errors and omissions need to be corrected as quickly and equitably as possible, because hardship for the requisitioner and supplier may result from slow problem resolution or unwillingness to pay for buyer-generated mistakes. Of course, the reverse of these arguments applies to seller mistakes or omissions. Remember that the cost of correcting a mistake or omission tends to increase rapidly over time and that fairness should be the guiding principle in covering extra costs incurred by the innocent party, whether it be the buyer or seller.

Price Not Specified
A special case exists when both parties have not agreed on a price in order to form a contract or have forgotten to specify a price. (Agreement on price is not necessary in order to form a contract.) Buyer and seller may agree on price later as long as two requirements are met. First, the parties must demonstrate their intent to contract by their writings or actions. Second, there must be some reasonably certain basis for fixing a price in the event of later conflict. Unless a contract contains specific language to the contrary, the UCC will presume an intent to contract even though the price is left open.

Rework and Damage
It is sound practice to involve the supplier in the decision to rework or to correct the situation and to obtain the supplier's agreement that the cost of correction or rework is reasonable. The damages incurred may be incidental, consequential, liquidated, or general.

Incidental damages are expenses associated with correcting the breach or attempting to rework or use nonconforming goods.

Consequential damages are indirect losses resulting as a consequence of the breach. They may include lost sales, the closing of a business, or personal injury, and can be very high. Sellers go to great length to disclaim responsibility for consequential damages.

Liquidated damages (Section 2-718 of the UCC) provides for prior agreement on an amount whereby damages will be liquidated in the event of breach. The amount must be reasonable in view of the potential for damages and the relative difficulty in presenting the case.

General damages include any damages resulting from requirements and needs of which the seller reasonably should have had knowledge at the time of contracting and that reasonably could not have been cured by the usual remedies.

Breach of Contract by the Buyer

There are at least four ways in which a buyer can breach a contract.

1. *Wrongful rejection.* Buyers must have a valid reason for rejecting goods delivered under a contract. They cannot make a contract and change their minds later.

2. *Wrongful revocation of acceptance.* The UCC distinguishes between original acceptance of goods and a later revocation of acceptance. Goods may be rejected for cause upon original receipt, but if they are accepted, the buyer cannot reject them later. A buyer can, however, revoke the acceptance if latent defects that substantially impair the value of the goods are discovered later. The seller must be notified and given an opportunity to correct the defect.

3. *Repudiation.* If the buyer advises (in advance of the time when seller performance is due) that he or she will not honor the contract, this is anticipatory repudiation and is a breach.

4. *Failure to tender payment due before delivery.* If agreed-upon buyer prepayments are not made, or if a buyer contracts for goods to be delivered COD and, upon delivery, will not pay for them, the buyer will have breached the contract.

Seller Recoveries for Buyer Breach

The seller can recover for buyer breach in the form of (1) resale damages; (2) market damages; (3) lost profits; and (4) contract price recovery.

Resale damages. These include the costs of selling the goods elsewhere.

Market damages. If, between the time of buyer breach and seller resale of the goods, the market value of the goods falls, the seller may seek the difference as damages.

Lost profits. In any suit for damages, the seller may include profits, up to the full amount, that were lost due to buyer breach.

Contract price recovery. If, after reasonable and diligent effort, the goods cannot be sold elsewhere, the seller may claim against the buyer for the whole contract price, less any net salvage value.

Contract Conclusion or Modification Options

There are many ways in which a contract may be concluded or changed. The most desirable and most frequent contract conclusion is where both parties have performed as agreed and both are fully satisfied.

Acceptance. According to law a buyer is considered to have accepted goods when, after a reasonable time for inspection the buyer:

- Advises the seller the goods are acceptable or that, if they are nonconforming, he or she will accept them anyway.
- Does not effectively reject the goods.
- Acts in a way inconsistent with seller ownership of the goods.

Acceptance of a part of any commercial unit constitutes acceptance of the whole unit. After acceptance, a buyer must pay for the goods as agreed and bear the burden of proving any nonconformity discovered later.

Revocation of acceptance. If, after goods are accepted, defects are discovered, the buyer may revoke acceptance under the following rules:

1. The defects later discovered must substantially impair the value of the goods to the buyer and must be true latent defects (i.e., not so obvious that they should have been discovered in the normal receiving inspection).

2. Revocation of acceptance can be based on a seller's failure to cure nonconforming goods, known upon receiving inspection to be nonconforming, but accepted based on reasonable expectation that the seller would cure.

Modification. A contract may be modified by agreement of the parties provided that the duties or obligations modified have not yet been performed and that the modification is equitable in view of circumstances not anticipated when the contract was originally formed. Under the UCC, additional consideration is not required to make a modification binding, but it must not have been based on fraud, duress, or any other violation of good faith.

Waiver. A party may waive its rights under an executory contract. A waiver clause should be specific and written so that the waiver of one right will not serve as a waiver of others. Care should also be taken to ensure that a waiver is not interpreted as ongoing and extending to other, similar agreements. Waivers may be retracted by reasonable notification unless the retraction would work an injustice on a party by virtue of the party's reliance on the waiver.

Rescission. Both parties may agree to rescind a contract before performance takes place. This will have the effect of setting the contract aside, with each party discharging obligations to the other.

Rejection. Within a reasonable time after defects are discovered (or should have been discovered with normal inspection), the seller must be notified of the defects and the buyer's rejection and be given reasonable time and opportunity to cure the nonconformity. Otherwise, the buyer will be barred from remedy. Furthermore, the buyer must not have waived his or her right to inspect.

Cancelation. Cancelation differs from termination in that it implies cause and does not excuse the "causing" party from damages resulting from its failure to perform. Contracts may be canceled for the following reasons:

1. *For cause or default.* This typically results in a cancelation; the implication is that of breach.

2. *For convenience.* The UCC does not provide a right to cancel for convenience. Therefore, any such right must result from agreement between the parties and is usually contained in a termination for convenience clause. Still, the code requires that all such actions be fair and taken in good faith. For example, terminations for convenience usually involve payment for executed performance and may include profits for the whole contract.

    Note that the government, in accordance with regulations, can terminate a contract for its convenience at any time, with or without cause. Furthermore, in government contracts, sellers cannot realize profits on that portion of the contract not performed.

3. *Commercial impracticability.* Under the UCC, if two parties enter into a contract and it later becomes "commercially impracticable" for one party to perform, that party will be excused from performance. However, the code states that sellers losing money on a deal will not make it commercially impracticable. Accordingly, performance will not be excused due to events such as shifts in markets, changes in costs, or others that should reasonably be foreseeable by business people; but events such as war, natural disaster, shortage of raw materials, or others that change the essential nature of the performance would probably be acceptable.

    To use this defense, a seller must within a reasonable time notify the buyer of pending nonperformance and, if the seller is not completely prevented from performing, must fairly allocate such performance as remains available among involved buyers. Failure by the seller to satisfy these requirements would prevent its being excused from performance under the concept of commercial impracticability.

After notification and a proposal for allocation, the buyer may accept the allocation or terminate the contract entirely. If the buyer fails to so modify the contract within a reasonable time (not to exceed 30 days) the contract will lapse with respect to any future performance.

4.   *Excusable delays and force-majeure.* The parties may agree upon the length and nature of excusable delays in performance. One common example of such an agreement is the force-majeure clause, which excuses a failure to perform when the failure is caused by unpredictable events over which the seller has no control, such as war, insurrection, or "acts of God" such as natural disasters. Generally, requirements for performance will be delayed until the force-majeure event is over and will then resume where they stopped.

Sellers often attempt to extend force-majeure clauses to include such arguments as transportation difficulties, failure of their suppliers to deliver, raw material shortages, and labor disruptions, over which the seller has limited or no control. Buyers should not allow the clause to become so broad that sellers are excused for events they could have influenced. Further, it is sound practice to trade additions to the force-majeure clause for concessions of value to the buyer.

Termination. Termination occurs when a party exercising a power created by agreement or law ends a contract for reasons other than breach. Upon termination, all executory obligations are discharged, but rights or obligations based on prior performance or breach survive.

Dispute Settlement

Disputes in purchasing most often occur because of differences in expectations between a buyer and seller, and only rarely because one party attempts to take advantage of the other. Accordingly, business people usually attempt to settle disputes in the simplest, least costly manner and with a view toward fair resolution rather than toward penalizing the other party. Dispute settlement methods include the following:

Renegotiation. The parties may place all their differences on the table and renegotiate the contract.

Alternate Dispute Resolution (ADR). Alternative dispute resolution provisions can be included in a contract to require both buyer and seller to follow a prescribed set of resolution attempts not unlike grievance procedures in a union contract. For example, if a buyer and a sales representative cannot resolve their differences, the ADR requirement may be that the purchasing manager and sales manager have to try it next, with vice presidents next and, ultimately, the presidents or chief executive officers of each organization. Only when this process fails can arbitration or litigation be used as the last resort.

Mediation. The parties, if unable to agree on settlement, may enlist the aid of a mediator who will listen to and question each side in an attempt to lead them to settlement. A mediator does not make the decision and the parties do not relinquish their legal rights.

Arbitration. In arbitration, the parties present their case to one or more arbitrators who, after hearing the evidence, will decide how it should be settled. The decision is binding on the parties and is enforceable through the courts. Unlike litigation, arbitration cases are private. They are also less costly, can be handled in less time, and do not involve evidentiary rules, discovery, or appeals.

Reformation. If, through fraud or mutual mistake, a written contract fails to express the true agreement or intentions of the parties, the courts may permit remedy by reformation of that contract.

Litigation. If unable to resolve the matter otherwise, a party may sue for damages suffered as a result of contract breach. The matter will then be decided in the courts according to the law. The process of suing and defending against a suit is referred to as litigation.

KEY POINTS

1. Purchasers are granted authority to commit their organizations to the purchases of goods, services, and equipment. Normally, such authority is extended to a specific spending limit.

2. A purchaser may be personally liable for commitments made.

3. There are many laws and regulations with respect to the supply field. Some of the better known laws include the Sherman Anti-Trust Act, the Clayton Act, the Robinson-Patman Act, and the Uniform Commercial Code.

4. A proper contract must meet five essential elements: (1) offer and acceptance, (2) consideration, (3) competent parties, (4) legality of purpose, and (5) genuine assent.

5. Contracts may be oral or written. Generally, contracts over $500 should be written.

6. Strategic, single source, and large contracts require special legal care.

7. Special clauses dealing with patents, copyrights, liability, suspension, force-majeure, comity, estoppel, assignability, insurance, and indemnification require legal assistance and review.

8. Warranties can be expressed or implied.

9. Suppliers in financial difficulties represent a major challenge, particularly when they have gone bankrupt.

10. Errors and omissions by both buyer and seller need to be dealt with equitably.

11. A buyer can breach a contract by a wrongful rejection, wrongful revocation of acceptance, repudiation, or failure to tender payment before delivery.

12. Contract conclusion or modification options include acceptance, modification, waiver, rescission, rejection, cancellation, and termination.

13. Dispute settlement options include renegotiation, alternate dispute resolution, mediation, arbitration, reformation, and litigation.

BIBLIOGRAPHY

Baker, R. Jerry, Lee A. Buddress, and Robert S. Kuehne. *Policy and Procedure Manual for Purchasing and Materials Control.* 2nd ed. Englewood Cliffs, NJ: Prentice Hall, 1993.

Burt, David N. and Michael F. Doyle. *The American Keiretsu.* Homewood, IL: Irwin Professional Publishing, 1993.

Burt, David N., Warren E. Norquist, and Jimmy Anklesaria. *Zero-Base Pricing™: Achieving World-Class Competitiveness through Reduced All-in-Costs.* Chicago: Probus, 1990.

Crosby, Philip B. *Quality without Tears: The Art of Hassle Free Management.* New York: McGraw-Hill, 1989.

Dobler, Donald W., David N. Burt, and Lamar Lee, Jr. *Purchasing and Materials Management: Text and Cases.* 5th ed. New York: McGraw-Hill, 1990.

Fearon, Harold E., Donald W. Dobler and Kenneth H. Killen. *The Purchasing Handbook.* 5th ed. New York: McGraw-Hill, 1993.

Fisher, Roger, William Ury and Bruce Patton. *Getting to Yes— Negotiating Agreement without Giving In.* 2nd ed. New York: Penquin Books, 1991.

Ford, William Obie. *Purchasing Management Guide to Selecting Suppliers.* Englewood Cliffs, NJ: Prentice Hall, 1993.

Gitlow, Howard S. and Shelly J. Gitlow. *The Deming Guide to Quality and Competitive Position.* Englewood Cliffs, NJ: Prentice Hall, 1987.

Heinritz, Stuart F., Paul V. Farrell, Larry Giunipero, and Michael G. Kolchin. *Purchasing: Principles and Applications*. 8th ed. Englewood Cliffs, NJ: Prentice Hall, 1991.

Killen, Kenneth M. and Robert L. Janson. *Purchasing Manager's Guide to Model Letters, Memos, and Forms*. Englewood Cliffs, NJ: Prentice Hall, 1991.

King, Donald B. and James J. Ritterskamp. *Purchasing Manager's Desk Book of Purchasing Law*. 2nd ed. Englewood Cliffs, NJ: Prentice Hall, 1993.

Leenders, Michiel R. and David L. Blenkhorn. *Reverse Marketing: The New Buyer-Supplier Relationship*. New York: The Free Press, 1988.

Leenders, Michiel R. and Harold E. Fearon. *Purchasing and Materials Management*. 10th ed. Homewood, IL: Irwin, 1993.

Moody, Patricia E. *Breakthrough Partnering: Creating a Collective Enterprise Advantage*. Essex Junction, VT: Oliver Wight, 1993

Nierenberg, Gerard I. *The Complete Negotiator*. New York: Berkley, 1991.

Pooler, Victor H. *Global Purchasing: Reaching for the World*. New York: Van Nostrand Reinhold, 1991.

Reilly, Norman B. *Quality: What Makes It Happen?* New York: Van Nostrand Reinhold, 1994.

Sherman, Stanley N. *Contract Management: Post Award*. Gaithersburg, MD: Wordcrafters Publications, 1987.

Sherman, Stanley N. *Government Procurement Management*. 3rd ed. Gaithersburg, MD: Wordcrafters Publications, 1991.

Walton, Mary. *Deming Management at Work*. New York: Perigee, 1991.

INDEX

A

ABC analysis, 21
Acceptance, 235
Acknowledgment, 215
Activity-based costing, 135
Ad valorem duties, 65
Adequacy of competition, 68
Administrative Procedures Act, 174
African-American Business
    Association, 76
Agency shop, 84
Allowable costs, 211
Alternate dispute resolution, 239
American Institute of Electrical
    Engineers, 41
American Institute of Mining and
    Metallurgical Engineers, 41
American Institute of Scrap Iron and
    Steel, 41
American Keiretsu, The, 14
American Mirrex Corporation, 132
American Society for Testing
    Materials, 41
American Society of Mechanical
    Engineers, 41
American Standards Association, 41
Annual work plan, 188
Antitrust, 223
Apparent authority, 206
Apparent scope, 207
Arbitration, 239
Armed Services Procurement Act, 213
Asian Business Association, 76
Assignability provisions, 224
Audit trail, 94
Automatic payment, 184

B

Baldrige Award, 45
Barclift, Lt. Jane A., 67
Battle of the forms, 216
Benchmarking, 13, 95
Better value, 94
Bid bonds, 104, 105
Bid log, 103
Bid or negotiate?, 99
Bid or quotation, 43
Bid solicitation concepts, 91
Bilateral contract, 217
Bills of exchange, 64
Binding letters of intent, 171
Blanket orders, 22, 163
Board of Contract Appeals, 174
Boilerplate, 225
Bonds, 105
Bottom line, 99
Breach of contract by the buyer, 234
Burlington Northern Railroad, 47
Burt, David N., 14
Business Research Services, Inc., 76
Buy America Act, 212
Buyer-planner, 55
Buying production or service
    capabilities, 129

C

Canadian Engineering Standards
    Association, 41
Cancellation, 237
Cancellation of solicitations, 99
Captive lease company, 121
Certified suppliers, 66
Chamber of Commerce, 74
Change orders, 192

Other books of interest to you from Irwin Professional Publishing

MANAGING STRESS
Keeping Calm Under Fire
Barbara J. Braham

As part of the *Briefcase Books Series,* this guide provides the author's CALM acronym for coping with stress and offers tips to help you learn to say "no." This *Briefcase Book* is an easy read with several guidelines for applying the techniques of stress management to life.
ISBN: 1-55623-855-X

BUSINESS NEGOTIATION BASICS
Peter Economy

This *Briefcase Book* is an easy-to-understand guide to business negotiating. You'll find seven basic techniques that guide you step-by-step through the negotiating process.
ISBN: 1-55623-841-X

MANAGING GLOBALLY
A Complete Guide to Competing Worldwide
Carl A. Nelson

This results-oriented book pioneers global strategic management, a process that holds the key to organizational success and survival in the new economic age. This how-to workbook includes practical checklists, flow charts, and matrices, making it possible for an organization to develop an effective global strategy.
ISBN: 0-7863-0121-X

GLOBAL PURCHASING
How to Buy Goods and Services in Foreign Markets
Thomas K. Hickman and William M. Hickman, Jr.

This valuable guide will teach you how to locate and evaluate foreign suppliers, negotiate purchases of goods and services, and arrange for shipment with maximum profitability and minimal risk.
ISBN: 1-55623-416-3

Available in fine bookstores and libraries everywhere!